Man on the Moon

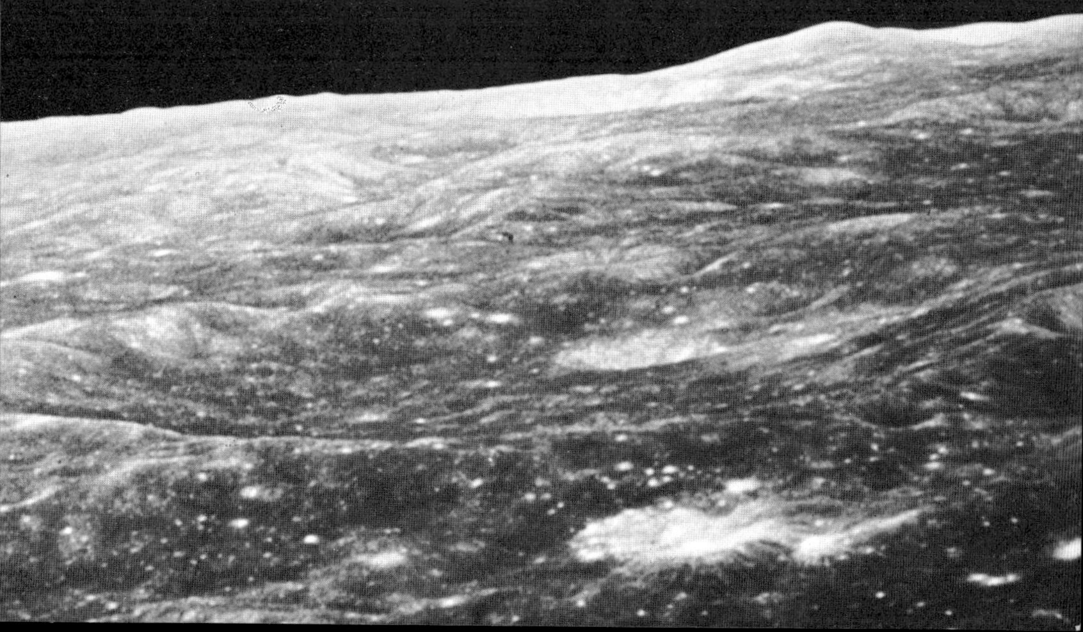

Frontispiece *Apollo 8 earth view. This view of the rising earth greeted the Apollo 8 astronauts as they came from behind the moon after the lunar orbit insertion burn.*

John M. Mansfield Constable London

First published in 1969 by **Constable & Company Ltd**
10 Orange Street, London WC2. Book design: Graham Bishop. Printed in
Great Britain at The Curwen Press Ltd, Plaistow E13
SBN 09 456600 3

To Fiona

Contents

Abbreviations

AMU	Astronaut Manoeuvring Unit
BIG	Biological Isolation Garment
Cap Com	Capsule Communicator
CM	Command Module
CMP	Command Module Pilot
CSM	Command Service Module
DOI	Descent Orbit Insertion
EDT	Eastern Daylight Time
EMU	Extravehicular Mobility Unit
EOR	Earth Orbiting Rendezvous method
EVA	Extravehicular Activity
ICBM	Intercontinental Ballistic Missile
IRBM	Intermediate Range Ballistic Missile
LEM	Lunar Excursion Module
LM	Lunar Module
LMP	Lunar Module Pilot
LOI	Lunar Orbit Insertion
LOR	Lunar Orbiting Rendezvous method
LRL	Lunar Receiving Laboratory
LSR	Lunar Surface Rendezvous method
MDS	Malfunction Detection System
MESA	Module Equipment Stowage Area
NASA	National Aeronautics and Space Administration
NP	Neutral Point (precise location of the exact balance between the gravitational fields of earth and moon)
PLSS	Portable Life Support System
SLMA	Spacecraft Lunar Module Adapter
SM	Service Module
SPS	Service Propulsion System
TEI	Transearth Insertion
TLI	Translunar Insertion
USAF	United States Air Force
VAB	Vertical Assembly Building
VFR	Verein für Raumschiffahrt (Society for Space Travel)

Acknowledgements

In writing this book I have made use of innumerable sources, many of which are acknowledged in the text. I should like to express gratitude for quotations to the following: the BBC, *The Times*, *The Daily Telegraph*, *Daily Express*, *The Guardian*, *The Observer*, *Life*, *Newsweek*, *Science Journal*, *The Economist*, *The Sunday Times*, and *Aviation Week*.

Illustrations

Illustrations

Illustrations: Copyright credits

Illustrations: Copyright credits

North American Rockwell **180**
North American Rockwell **181**
North American Rockwell **181**
USIS **183**
North American Rockwell **184**
North American Rockwell **185**
North American Rockwell **185**
Novosti **186**
NASA **217**
NASA **220–1**
Associated Press (bottom) **225**
Grumman, **242–5**
USIS **245**
USIS **247**

Preface

'One small step for man, one giant leap for mankind.'

'Man on the moon' – only a few years ago the idea of a man walking on the moon was regarded by many as an improbable dream. On 21 July 1969, Apollo 11 converted that dream into an historical fact. The space conquests of this year have happened so quickly that it is still difficult to grasp their full significance. The pageant of dramatic events in the exploration of space has achieved such a whirlwind pace that many, including myself, find it hard to believe that they have really taken place at all. The space epics of the cinema had prepared us for the sight of Neil Armstrong clambering down to the lunar surface – but the problem was to convince oneself that, at last, 'this was for real!' – and not yet another Hollywood spectacular.

One year ago the publishers kindly invited me to write a book to commemorate man's landing on the moon. At that time no one knew when the final moon shot would be, and no one knew whether the book would climax with an American success story, or with one of the greatest disasters in history. Today the choice of the title seems naïvely obvious – yet only a few weeks ago it could have been the ironic title of a tragedy. Even as it is, the story which leads up to man's conquest of the moon is by no means just a technological account of scientific progress: it is very much a human drama of triumph and death, success and tragedy, ending with the fulfilment of an impossible dream.

'This is the greatest week in the history of the world since the creation.' That was how President Nixon described the third week of July 1969, as he personally greeted the astronauts Armstrong, Aldrin and Collins on their safe return to earth. This book describes the epic events of that week in detail. It also looks forward, beyond man's first step into the unknown, revealing the probable future of interplanetary travel. It reviews how science fiction has been converted into fact, and traces the fascinating history of man's growing interest in his universe from the earliest times.

From Leonardo da Vinci to the Wright brothers is a brief span of 400 years, but those years only encompass man's learning to fly in air. Shakespeare thought our atmosphere extended to the stars and beyond; he expressed man's everlasting wonder at the heavens in 'Hamlet' – 'this most excellent canopy the air, look you, this brave o'erhanging firmament, this majestical roof fretted with golden fire.'

Today man has entered that 'brave o'erhanging firmament' – 'A magnificent picture was spread out before me. I could see both the steady brilliance of the stars scattered over a background of black velvet, and at the same time, the surface of the earth – just as though I had been floating over a great coloured map.' That was how Colonel Leonov, the first man to walk in space, described what he saw when he floated away from his spacecraft. Only four years later the astronaut 'Buzz'

Aldrin muttered 'Beautiful, beautiful' as he stepped onto the moon's Sea of Tranquility – 'A magnificent desolation'. Man had begun to colonise other worlds.

Normally several months elapse between the delivery of a manuscript and the arrival of the first bound copies. To get this book ready within weeks of the moon landing has been quite a feat in terms of both printing and publishing. Many parts of it were prepared well in advance of July, and I should like to thank Constable and Miles Huddleston in particular, for not only their patience and encouragement, but also for their very hard work without which the book would never have been finished. I am grateful to The Curwen Press and the binders, William Brendon, for their enthusiastic co-operation and the speed and efficiency of production.

In compiling and writing this book I have received enormous assistance and advice from many friends and colleagues. In particular I would like to thank James Burke: he is the only man in Britain who needs no sleep, judging by his endurance feats as he reported day and night to British television audiences on the flight of Apollo 11. His command of the subject, combined with his wit, enthusiasm, and prodigious memory probably gave BBC viewers a better insight into the real achievements of Apollo than any other television audience on earth. His assistance, in matters of both content and style, has radically improved this book. I should also like to thank Patrick Moore for encouraging me to venture into print on this subject in the first place: he partnered James Burke in the Apollo TV endurance trials and has written countless excellent books and articles about the moon.

I am very grateful to Dr Wernher von Braun and Dr Ernst Stuhlinger of the Marshall Space Flight Centre for giving me so much of their valuable time, and answering my many questions with unflagging enthusiasm. My thanks are also due to Professor George Mueller, for permission to quote some of his predictions in Chapter 14, and to Lord Russell for allowing me to quote from his article 'Why man should keep away from the Moon', published in the London *Times*, on the eve of the Apollo 11 moon-launch.

I am indebted to the British Broadcasting Corporation, as a member of its staff, for being granted permission to write this book.

I should also like to thank Stuart Harris, Bob Toner, and Joy Tapply. Bob Toner helped me with the research for the historical portion of the book, and Stuart Harris helped me evaluate and write the post-war story. With Joy Tapply, they gave me invaluable advice, and without all their kind assistance this book could not have been written.

August 1969 John M. Mansfield

Introduction

Soon as the evening shades prevail,
The moon takes up the wondrous tale,
And nightly to the listening earth
Repeats the story of her birth.

(Joseph Addison, 1672–1719)

In the year 1969 man at last learnt the real story of that birth. From time immemorial men have gazed into the sky and pondered, theorised, even worshipped 'the silver ornament of night' (Barnabe Barnes, 1598): 1969 marks the final transition from mankind being moon-struck, yet firmly earth-bound, to actually setting foot on the moon.

Man's conquest of the moon is a history of extremes. Everything about it is in superlatives – from the cost, the effort, the danger, the power, and the distance, to the time-scale involved. Perhaps the most remarkable extreme about it is the relatively short time in years that it has taken for man to change from a primitive moon worshipper to an interplanetary explorer. It is the story of man's greatest technological adventure, of both the biggest prestige battle of all time, and the strangest ever progression from fiction to fact. Even half-way through the twentieth century, in 1950, very few people on earth seriously thought that within twenty years man would be in a position to land on the moon. As late as 1956, even the Astronomer Royal, Professor Woolley, was quoted in the Press, saying: 'All this talk about space travel is utter bilge'. Yet today, although the names Sputnik, Gagarin, Eagle, Columbia and Apollo have become familiar in every language on earth, the extraordinary story that lies behind those peaks of human achievement still remains largely unknown. Many critics comment on the controversy about whether man, in the face of growing food shortages and the population explosion, could afford his attempts to get to the moon: few of them realise that, although the American expenditure on space is colossal, it amounts to only two per cent of the United States' Gross National Product, as is explained later in this book. Also, that 'wasted' money has enabled almost unbelievable advances to take place in computers, medicine and a myriad other frontiers of science.

Doubtless an international mood of anti-climax will gradually spread after the moon landing; it will probably be greater than that felt after the first ascent of Mount Everest. However I hope this book will do something to counteract that sentiment. The moon is no final goal, but rather, an important stepping-stone in a series of extra-terrestrial challenges. It should be regarded as the beginning rather than the end of an era. What impressed me most in this context was the Vertical Assembly Building (VAB) at Cape Kennedy, the largest building of its kind in the world: it is 525 feet high, 716 feet long, and, without air-conditioning, clouds would form inside its roof. It contains four separate bays for the vertical assembly of the gigantic moon rockets, the Saturn V's. It is covered by more than a million square feet of glistening aluminium, and its foundations rest on 4,225 steel pilings, each 16 inches in diameter, driven from 150 to 170 feet to bedrock. The skeletal structure alone contains 57,000 tons of steel. Such statistics are indigestible, but looking down inside it at the Saturn V destined to lift Apollo 9, I was impressed not so much by the almost incredible size of the building, as by its permanence.

And then – as James Burke said to me – we were preparing a special report on Apollo 8 at the time for BBC 'Tomorrow's World' – 'it's like a bus station'. And it was. The VAB is no overgrown Nissen hut designed for some temporary function: it is built to last, and a National Aeronautics and Space Administration (NASA) official coolly pointed out that it was conceived to house launch vehicles which not only are not on the drawing boards, but have not even left the doodle pads of future inventors.

Adjoining the VAB are the four firing rooms, each with forty feet high windows looking down the crawler pad to the launch site three and a half miles away. Here again it was not the rows of consoles and displays, nor the hundreds of men staffing each room that emphasised the staggering scale of the enterprise, nor its modernistic nature – it was details like the venetian blinds shielding each firing room against the intense Florida sun. The blinds work on a vertical axis, each blade being about 3 inches thick, 3 feet wide, and 40 feet tall; they are electrically powered following the commands of photoelectric cells in order to maintain a constant level of illumination in the rooms in daylight hours. Even millionaires would think twice before installing such giant extravaganzas – yet in the firing rooms they somehow seemed appropriate, and by saying that I am not implying a budget-conscious criticism of NASA: monster automatic venetian blinds may seem a curious absurdity to select from all the gadgetry of Cape Kennedy, yet as examples of ultra perfectionism, they are in many ways typical of the whole American space effort.

The moon is nothing
But a circumambulatory aphrodisiac
Divinely subsidized to provoke the world
Into a rising birth-rate.

So wrote Christopher Fry in *The Lady's not for Burning*, but by the time the American astronauts and Russian cosmonauts have finished with the moon, it may not even be remembered as the prime mover of the oceans, the female menstrual cycle, and human madness. Yet this target is in no ordinary satellite orbit: today the moon is receding very slowly away from the earth. However, some astronomers, after complex calculations, have worked out that many years from now it will start coming back and eventually get as close as 11,000 miles. At that point earth's gravity will probably cause it to disintegrate. Whatever happens then is open to speculation. Fortunately it is unlikely to happen for 30,000 million years.

Until a detailed analysis of samples of lunar soil has been completed, the exact origin of the moon remains a mystery. The most popular theory still taught in many schools is that the moon was originally part of the earth, flung into space from the Pacific Ocean area by an interaction of the sun and the earth. Another is that the moon was originally a satellite of the sun in a completely different orbit from today, and, passing close to the earth, was attracted by earth gravity into an earth orbit.

O Moon, when I gaze on thy beautiful face,
Careering along through the boundaries of space,
The thought has often come into my mind
If I ever shall see thy glorious behind.

Those lines were written by a housemaid poet (quoted by Robert Ross in *Academy*): ten years ago in October 1959 the Russian Lunik 3 succeeded in photographing that 'glorious behind', and at Christmas 1968 the Apollo astronauts, Borman, Lovell and Anders, saw it with their own eyes. Today we remember the fact of that first manned moon orbit, and we tend to forget the very real dangers involved, and the courage of the first explorers of outer space. Yet in November 1968 the eminent British astronomer, Sir Bernard Lovell, was quoted by the London *Evening News* as condemning the proposed flight of Apollo 8 as 'just bloody silly'. He said: 'There is a very good chance that these three men will never return to earth. Or if they do they will be dead.'

'They could expire in a lunar orbit they cannot get themselves out of. They could be killed by radiation . . . or by the force of re-entry into the earth's atmosphere. The thought of this flight sickens me.'

Against that climate of criticism I talked to the astronauts only a few days before their historic journey. Their attitude to the flight was that of supremely confident engineers, men satisfied that every aspect of the mission had been checked, rechecked, and checked again. There was no melodrama. Frank Borman said: 'What everybody looks forward to right now is to hear those bolts go.' (The Saturn V is held down on its launch pad for 8·9 seconds after the ignition of the giant first stage of the rocket to enable it to develop full power for lift-off.) Jim Lovell added: 'It's really an anti-climactic because the apprehension towards the flight is now – you know, the last month or so . . . . Once you're in the spacecraft, once the hatch is closed and the engines start up and those retaining arms go down, then it's either gonna go or it's not gonna go.'

Grissom, White and Chaffee died during a rehearsal of such a launch procedure, and Komarov, the Russian cosmonaut was killed during re-entry from a space flight. Their deaths underline the very real perils of man's excursions into space: many say that unmanned spacecraft could carry out equally good research at less cost and without the risk to human life. And when man has landed, what will he find? Dr Wernher von Braun, one of the men most responsible for both the German technical success with the V-2, and the recent developments in the American space programme, was once asked what the Americans would find when they landed.

He replied, 'Russians'.

Ever since President Kennedy committed the United States to landing an American on the moon by 1970, there has been, as far as many of the protagonists are concerned, a space race.

The Russians deny that they are involved in anything so degrading as a race – but whether they like it or not, most of the world considers that they are. This book traces the story of that race. It also attempts to go further – the space race element encompasses little more than ten years of hectic human endeavour: this book will try to put the space race into perspective, seeing it against the background of man's growing technological ability. Impossible dreams of the past are today becoming accepted reality, and tomorrow may lead man further than just the moon, Mars, or even our galaxy, to cross what we think of now as the limits of outer space.

'The Moon is like a Chariot Wheel' Ancient concepts of the moon

It is impossible to say exactly when the moon was first consciously observed and studied by man. The early cave-paintings reveal his interests as very down-to-earth; nearly all show his fight for survival, often in epic tableaux of the hunt. There is little evidence that he paid much attention to the night sky, and there are no pictures of either the moon or the stars. However it seems certain that early man soon appreciated the life-giving significance of the sun, and began to worship it. Stonehenge is often mentioned as one of the first tangible signs of man becoming conscious of the predictable behaviour of the sun and the moon, and early graves have been unearthed to reveal bones marked with primitive indications of the various phases of the moon.

Sir Leonard Woolley's excavations of cities dating back to 4000 BC in the Tigris-Euphrates valley uncovered some of the earliest records of man's first interest in the universe: at that time the sun had been demoted from the top rank of the celestial hierarchy, and was worshipped as Utu, the son of Nanna, the moon god, and chief astral deity.

Nanna, according to the Sumerians, plied across the sky in a quffah (a circular type of boat used on the Euphrates), accompanied by his retinue of stars and planets. It was only in the Hebrew cosmogony, with the sun restored as the major luminary, that the moon underwent a sex change to emerge and remain female for virtually the whole of classical mythology.

Clay tablets dating from about 3800 BC show that astronomy had already become an established practice. Man's innate conceit, or some might say perception, soon persuaded him that his life was inextricably involved with the movement of the stars, and astronomical observation became as confused with astrology as it is in today's daily horoscopes.

By 3000 BC the Babylonians were able to calculate the movement of the sun, moon, and planets in advance. They worked out lunar calendars, starting their year with the first new moon following the spring equinox. Inevitably this meant an occasional calendar re-adjustment, and the King of Babylonia ordered an extra month to be inserted in 2000 BC – so profiting from an extra month of taxes. It was at about this time that the signs of the Zodiac were applied to the various constellations.

In about 1500 BC the Babylonians finally settled on a calendar containing months and years which did not require rearranging from time to time. By 1300 BC the Egyptians had catalogued several constellations and individual stars, and established a solar calendar of 365 days, a marked advance on the 30 or 29 days-in-a-month lunar calendar. They also invented the sun-dial, and calculated that Venus and Mercury were closer to the sun than earth, Mars, Jupiter, and Saturn. Some of their legends explained the waxing and waning of the moon as the action of a ferocious sow, which approached the moon god on the 15th of each month, and gnawed away until he disappeared, to be re-born for the next lunar cycle; during an eclipse the sow swallowed the moon god whole, and in a solar eclipse the sun was engulfed by a serpent.

The Egyptians developed a cult of sun worship at Heliopolis. Re was the sun god, and his deputy was called Thoth, the moon god. Thoth held sway over darkness, the underworld, and all the enemies of Re; perhaps ironically Thoth was also honoured as patron of the sciences. The Egyptians had another moon god, called Khonsu – the moon has always been regarded as a celestial time-piece, and Khonsu, very aptly, was also the god of Time. Even today the natives of New Guinea reckon months by the moon, and some of them throw stones and spears at it in order to speed it through the sky and hasten the return of friends, away for a year at a time, working on the tobacco plantations. The Malays even hurl ashes and spit at the moon, in an attempt to extinguish it, for fear that its glow should throw some weak person into a fever.

In legend and mythology the moon often has a rather sinister role. In one German tale the man in the moon was a villager, caught in the act of stealing cabbages, and placed on the moon as a warning to others. On the island of Sylt he was believed to be a sheep-stealer, and in Polynesian legend he is known as a thief. The nursery tale of his being a man banished for gathering sticks on the Sabbath has parallels in many countries.

The Turks have a more romantic legend – to them the moon was a young bachelor, betrothed to the sun. Originally the sun had shone at night, with the moon taking the day shift, but the sun, a girl, was afraid of the dark and so she changed places with her lover.

In literature the moon has regularly figured as a symbol of either sorrow, magic, or love – and often as a mixture of all three. In *Astrophel and Stella* Sir Philip Sidney typified an attitude towards the moon which still colours our view today, however scientifically precocious we may be:

With how sad steps, O Moon, thou climbst the skies!
How silently, and with how wan a face

In *Paradise Lost* Milton virtually sums up the serene, magical powers of the moon:

Fairy elves,
Whose midnight revels, by a forest side,
Or fountain, some belated peasant sees
Or dreams he sees, while overhead the moon
Sits arbitress.

Even today much of our vocabulary for lunar studies emanates from mythology: to the Romans Luna, the moon, was the sister of Helios (Sol – the sun), and of Eos (Aurora – the Dawn). The word selenology, the study of the moon, comes from the Greek mythological figure called Selene, the goddess of the moon.

Diana the huntress was also identified with the moon, but, before or after setting, the moon was known as Hecate in classical mythology, and even as Phoebe the sister of Phoebus when her relationship as sister of the sun needed emphasis.

The ancient Greeks credited Thales of Miletus with founding science, mathematics and philosophy, but he had to go to Egypt to learn astronomy. He succeeded in predicting a solar eclipse, and when it occurred in 585 BC, it frightened the armies of Media and Lydia into making peace. Thales was what we would call today 'a flat-earther'; he considered the earth to be an island or raft in the middle of a sea domed by the stars.

One of his contemporaries, Anaximander, proclaimed that the moon was 'a circle 19 times as large as the earth; it is like a chariot wheel, the rim of which is hollow and full of fire, as that of the sun also is; it has one vent, like the nozzle of a pair of bellows; its eclipses depend upon the turning of the wheel'.

The first argument against the flat earth theory was put forward by Aristotle, and he used his observation of the moon to prove it. By that time the Greeks had realised that the moon did not shine by its own light, but reflected the rays of the sun.

The earth was a sphere because the moon was opposite the sun at a total eclipse, therefore the darkening was due to the shadow of the earth. The edge of that shadow always had a circular outline, whatever the position of the full moon; therefore the earth was the rotational figure of a circle – a sphere.

By the fourth century BC there were several conflicting theories: Parmenides of Elea argued that the sun, moon, and planets were arranged in bands around a spherical earth; others thought that the earth, moon and sun all revolved around a great ball of fire which constituted the centre of the universe.

Heraclides of Pontus (388–315 BC) explained the daily rotation of the stars by assuming that the earth turned on its axis. He also discovered that Mercury and Venus revolved around the sun. A few years later Aristarchus of Samos (310–230 BC) had the revolutionary idea for his time that the earth revolved around the sun: it took 2,000 years for that idea to be widely accepted, and even today the Flat Earth Society presents the most ingenious arguments against it. The concept of a spherical earth became embedded in Greek thought, and even Plato, at times an arch reactionary, accepted its truth: he believed that it was only man's 'weakness and sluggishness' that prevented him travelling upwards through the air.

In an essay about 'The Face on the orb of the Moon' Plutarch summed up what was already known about selenology, and added some fantasies of his own. In the form of a dialogue he suggested that the markings on the moon were caused by either defects in our own eyes, or reflections from earth. 'The moon contains deep places and chasms, and just as our earth is split by deep depressions, so the moon is gouged with great clefts and depths, containing water and dark air, unlit by the sun'. He felt the moon could easily be a dwelling place for spirits. He also proposed that the moon did not fall on the earth for the simple reason that it was moving around us, a concept which had to await fifteen centuries before being proved by Newton.

Plutarch died in AD 120, and that year a theoretical astronomer was born whose ideas were to remain unchallenged for the next fourteen hundred years. His name was Claudius Ptolemaeus, and his book, the *Almagest*, is an invaluable summary of the science of his time. He considered the earth to be a globe around which moved the sun, moon and planets, with the star universe at a much greater distance. Since the circle was regarded as the perfect form, nothing short of perfection could exist in the sky. However observation had shown that the planets did not describe perfect circular orbits: he therefore ingeniously suggested that each planet moves in a small circle, the centre of which revolves around the earth in a much larger, but equally perfect circle.

The rigid Ptolemaic concept of world order was to remain virtually unchallenged for 1,400 years. However, a contemporary of Ptolemaeus, Lucian of Samosata, could be called the first science fiction writer. Born in Asia Minor in AD 120 he wrote the *Vera Historia* (True History): it contained many elements of modern space-travel – both fact and fiction. He described a space flight, a landing on another world, a description of that world, and a return. A Greek satirist, Lucian went to great lengths to assure his readers that he was not telling the truth.

With fifty companions the narrator sets out on a voyage to the then uncharted Western Ocean. After passing through the Pillars of Hercules (the straits of Gibraltar) the ship is caught in a whirlwind and lifted into the sky. The wind fills the sails and they travel for seven days and nights; on the eighth day they see a large tract of land, like an island, round, smooth, and full of light. They have arrived on the moon. The moon is inhabited by the Hippogypi, whose warriors ride on three-headed vultures adorned with feathers 'bigger than the masts of a ship'. The astronauts are captured and brought before the King of the Moon, Endymion. They help him in an interplanetary war, and after visiting a colony on Venus, return to earth. In another story of space travel Lucian writes of the hero, Menippus, strapping both a vulture's wing and an eagle's wing to his back, and using Mount Olympus as his launch pad. After reaching the moon he decides to venture further, and in three days he reaches the stars, and the heavens themselves. Jupiter is angered by this intrusion into his domain, and orders the wings of Menippus to be cut off.

With the fall of the Roman Empire, Hellenic culture passed to the Arabs. The Greek and Latin texts were translated and studied. In the ninth century AD, Muhammad al-Batani calculated the precession of the equinoxes, and a hundred years later Ibn Junis recorded both solar and lunar eclipses.

The Arabs had the remarkable belief for that time that the earth's atmosphere thins out with altitude instead of filling the whole universe. An epic poem of this period may contain an even earlier record than Lucian's of man's first imagined venture into space, for it was based on ancient legends. It is called *Shah-Nama*, by the Persian poet Firdausi, and tells of a hero called Jamshid who was carried through the heavens on a throne borne by demons. In another part of the poem Kai-Kaus, the mythical king of Persia, is persuaded by a demon to attempt the conquest of the sky. He hangs legs of lamb from lances, attached to his throne, and then binds four young eagles to the throne. Once comfortably seated on his regal launch vehicle, the eagles hurl themselves into the air to reach the meat, and carry him aloft.

From Copernicus to Jules Verne The theoretical foundation is laid for the space age

'Of cold and human nature with the humidity of water rather than that of air' . . . such was the extent of scientific knowledge about the moon in the thirteenth century. The description was contained in a series of commentaries called *Tractatus de Sphaera*, which taught that the earth was the centre of the universe. The concept was held as being fundamental to all Christian doctrine, and when Copernicus published his arguments against the theory in 1543 he was regarded as a dangerous heretic. Copernicus bravely revived Aristarchus' theory that the earth and planets revolved around the sun. He clung however to the classical idea that the planets orbited the sun in perfect circles.

It is perhaps rather fitting that one of the greatest artists of the Renaissance should have solved a mystery of the moon by using his own meticulous powers of observation: Leonardo da Vinci, who died more than twenty years before Copernicus published his heresy, noted that during the crescent moon the dark portion can often be seen quite clearly. He was the first to explain this phenomenon as a reflection of light from the surface of the earth, and within a hundred years of his death the rebirth of interest in science and astronomy had revealed many of the fundamentals of space travel today.

By 1600 the Danish astronomer, Tycho Brahe, had measured the positions of the stars and planets with astonishing precision. He was not prepared to be damned as a heretic, but compromised on the Copernican theory by stating that Mercury and Venus, unlike the earth, orbited the sun – which in its turn orbited the earth.

The key to progress in astronomical theory was the invention of the telescope, but it took many years for the device not to be regarded as a piece of technical trickery giving the user optical deceptions. Galileo radically improved its design, obtaining magnifications of thirty times whereas previous models could only give an image three times life size. He discovered sunspots, and ruined his eyesight by peering at the sun. In 1610 he published his *Nuncius Siderius*, revealing that Jupiter has satellites just as the earth has the moon as a neighbour. Most philosophers of the time considered all the heavenly bodies to be perfectly round and smooth, but Galileo contradicted this, observing the moon to be as full of irregularities, mountains, and valleys as the earth itself. The cultured world was scandalised, but at the end of that year Johannes Kepler, a pupil of Brahe, came to Galileo's defence with a theoretical justification for his observations.

In 1604 Kepler had evolved the basic principles of the new science of optics, and by 1609 had discovered that a planet moves around the sun in an ellipse. He also found that the radius of this ellipse sweeps equal areas in equal times. These were two of his famous Laws of Planetary Motion. The third was published in 1619 – the square of a planet's sidereal period is proportional to the cube of its mean distance from the sun. The theory sounds daunting, but it means that the time a planet takes to orbit the sun is directly related to its distance from the sun. Meanwhile Galileo was even trying to calculate the height of some of the mountains, and the depth of the major craters. However his erudite defence of Copernicus resulted in his being brought before the Inquisition, and forced into a recantation of his views. In 1641, under house arrest, he died, blind.

Seven years before the death of Galileo, Kepler had published his *Somnium* (The dream), the first lunar guide book. In a pseudo-scientific manner he divided the moon into Subvolva, and Privolva, Subvolva being the part always visible to earth, and Privolva the unseen portion which the Russians were the first to look at in 1959. The *Somnium* involves spirits who can only travel at night: they transport humans to the moon under cover of the darkness caused by an eclipse. Because of the extreme contortions which any human would suffer on such an excursion to the moon, Kepler visualised him being put to sleep with his limbs arranged to absorb the shock. He thought that earth's atmosphere did not stretch as far as the moon, so suggested the application of a wet sponge to the astronaut's nostrils during any difficulty in breathing. He felt that the moon must be peopled, but that its inhabitants only emerged from their caves into the bright sunshine on rare occasions, preferring the gloom of the shadows. It was Kepler's work that Samuel Butler thought of when he wrote:

> . . . Th' Inhabitants of the Moon,
> Who when the Sun shines not at noon,
> Do live in Cellars underground,
> Of eight miles deep and eighty around.

A hundred years earlier Astolpho in *Orlando Furioso* had journeyed to the moon in a chariot drawn by four red horses; the moon was described as possessing most of the features of earth, with cities, towns and castles: in half a century science had made remarkable advances, but fiction had already established the moon as an ideal background for fantasy and dreams – and some of those dreams, the fiction of later writers like Jules Verne, were, in their turn, to be converted into the realities of travel to the moon today.

In spite of ecclesiastical anger at the new ideas of Copernicus and Galileo, even the clerics profited from the new interest in fictional astronomy. By 1638 the Bishop of Hereford, Francis Godwin, had written his book *The Man in the Moon*. He described how Domingo Gonzales, a young Spaniard marooned on St Helena, escapes by training wild geese to carry him away in a chairlike device, called 'the engine'. Unfortunately the geese decide to migrate to the moon, and fly him there in twelve days. The moon closely resembles the earth, covered by a large sea which covers 'three parts in foure, if not more'. It is a veritable Utopia, with many herbs, birds, beasts, and trees three times as high and five times as thick as those on earth. The return to earth takes less than nine days, and this – with a distinct parallel with today's *Apollo* missions – is explained by the earth having a greater attraction.

Bishops in the mid-1600's seemed fascinated by the moon, for another, John Wilkins, the Bishop of Chester, wrote *Discovery of a New World* on the same subject. It was intended to be taken as a serious essay. Wilkins was convinced that the main problem in attempting to travel to the moon was to lift the astronaut to a point between moon and earth where the earth's influence ended. He calculated that point to be approximately twenty miles above the earth's surface. Beyond that the rest of the journey would be simple, and since our bodies would be devoid of gravity, no further effort would need be exerted, and strangely enough no more food would be required.

Wilkins imagined wings being attached to the traveller's body, but he gave no details about whether the man himself would have to power them. He strongly believed the moon to be inhabited, and urged the British Government to acquire it for the nation.

In 1649 Cyrano de Bergerac devised perhaps the most ingenious method of all time for reaching the moon. He had observed that the dew rises heavenwards once exposed to the sun. In *Voyage to the Moon* he suggested tying a large number of flasks, filled with dew, around the waist. The hopeful astronaut should then stand in the morning sun until the dew evaporated and lifted him high above the earth. In this particular account de Bergerac admits that the method was somewhat unsuccessful – his traveller only managed to fly as far as Canada. In 1652 he published another work of science fiction entitled the *Comical History of the States and Empires of the Sun*. In this he almost outdoes Jules Verne in prophetic originality; he builds a flying machine, and jumps off a mountain with it, only to crash into the valley below. He leaves the machine, and on his return discovers soldiers attaching firecrackers to it. He jumps aboard at the very moment when they light the fuses, and the rockets lift the machine high into the sky. Once the fireworks are exhausted he begins to lose altitude, and falls – on to the moon. De Bergerac's powers of invention then suffer a temporary lapse, for the return journey is made under supernatural power.

De Bergerac's next idea was admittedly as impractical as his others, but almost foresaw modern jet propulsion. He builds another launch vehicle to take him to the sun and planets; he describes it as a box, three feet by six feet, on the top of which is balanced a twenty-sided glass figure: this Icosahedron focuses the rays of the sun into the box. The air in the box heats up and expands through tiny holes underneath, causing the whole contraption to rise. As the air becomes thinner, the propulsion system is less effective and, Cyrano continues, propelled solely by will-power.

Isaac Newton was born ten years before the publication of the *Comical History*, so it is just possible he read it as a young man. Certainly he was well acquainted with Kepler's work, for in his *Mathematical Principles of Natural Philosophy* (1687) he interpreted Kepler's Laws of Planetary Motion by his own universal laws of motion. He even invented calculus in order to find the derivation of Kepler's Laws. A brilliant mathematician, he explained the movements of the universe by his laws governing Motion, Inertia, Acceleration, and Gravity. Combining the Laws of Motion of both Kepler and Galileo he evolved his own theory for projectile orbit, explaining that if a cannon is placed on a high mountain and a cannon ball fired above a certain velocity, that cannon ball would remain in orbit and not fall back to the ground on a normal ballistic trajectory. His Law governing the Principle of Action and Reaction ('for every action there is a reaction') is the key to all rocketry, and, with his theory of orbit, it established the basis in theory for the Space Age. Ever since Newton the history of astronomy has been largely the working out of a universal system following his guide lines.

Three years after Newton published his *Mathematical Principles*, Gabriel Daniel's *Voyage to the World of Descartes* was printed. It is tempting to read into it today a spiritual rebellion against the newly discovered limiting force of gravity, for Daniel introduced the idea of soul travel. In this *Voyage* the soul separates from the body and goes in search of Descartes. It travels to the moon and beyond, describing the moon *en route* as closely resembling our earth. In 1703 David Russen had a more physical, albeit painful method of reaching the moon – in his *Voyage to the Moon* he proposed riding in a carriage fired by a large spring catapult. Two years later Daniel Defoe reviewed several legendary accounts of lunar missions in his book *The Consolidator*. The Consolidator itself was described as 'an engine in the shape of a chariot, on the backs of two vast bodies with extended wings, which spread about fifty yards in breadth, composed of feathers so neatly put together that no air could pass. . . .' Another flying chariot crops up in Samuel Brunt's *Voyage to Cacklogallinia*, part of which involves a trip to the moon in search of gold. The chariot is propelled by birds, and an interesting aspect of the flight is the author's notion that the air becomes less dense as the altitude increases. The birds, instead of flying straight up into space, rise slowly to accustom themselves to the changing atmospheric conditions.

By 1827 science fiction was becoming more sophisticated: in Joseph Atterly's *Voyage to the Moon* the spaceship is loaded with scientific equipment, and rises to the moon by means of an anti-gravitational material called Lunarium. Even Edgar Allen Poe dabbled in lunar fiction in his story of Hans Pfaal (1835). Hans takes off in a balloon to escape from his creditors, carrying with him an air condenser to enable him to breathe more easily in space. Hans eventually reached the moon and stayed there. Even at that time serious scientists still believed that there was life on the moon. One of the greatest astronomers of all time, Sir William Herschel, who discovered the planet Uranus, considered there were beings not only on the moon, but also the sun. He thought that inside the brilliant solar surface there could well be a region sufficiently cool to support life, and it was his son whose name became world news in 1836, in the Great Moon Hoax.

The Great Moon Hoax was perpetrated by a mischievous English journalist, Richard Locke, in 1836. It appeared as a genuine report in the *New York Sun*, under the arresting headline, 'Great Astronomical Discoveries Lately Made by Sir John Herschel at the Cape of Good Hope'. The articles appeared in serial form, and lasted a week, gaining credibility because they were alleged to contain facts submitted by Sir John to the supposedly august (but actually defunct) *Edinburgh Journal of Science*. The moon was described as inhabited by a strange variety of people, ranging from satyr-like creatures to semi-humans. The public's acceptance of anything scientific, plus the convincing manner in which the articles were released, made Locke's story so credible that it was universally accepted for several days. There was considerable public disappointment when it was eventually exposed, and the *Sun* admitted the hoax. Locke timed his hoax well, for the very next year its authenticity might have been doubted, after the publication of the first good map of the moon: in 1837, with the aid of a superb telescope in Berlin, Wilhelm Beer and Johann von Madler drew up an excellent lunar chart, complete with descriptions of the moon's surface.

The next major fictional voyage to the moon, although technically unfeasible was sufficiently plausible to become probably the greatest single stimulus for the pioneering of manned space flight in the history of imaginative literature. It was Jules Verne's *From the Earth to the Moon*, published in 1865, and its sequel *Around the Moon*.

Verne suggested firing a 20,000-lb. spacecraft at the moon, using a monster cannon, 900 feet long. He liaised with the Observatory of Cambridge to find the most suitable time for the launch. The projectile had a conical nose, and could accommodate three astronauts, a dog, and a mass of scientific equipment. The capsule, called 'The Columbiad', was lined inside with thick leather padding. Access was by means of 'a narrow aperture in the cone, secured by an aluminium plate'. The travellers could look out through four lens-shaped portholes with heavy metal lids. The provisions for the journey were remarkably sophisticated, and included chemically supplied air.

The tremendous acceleration of such a projectile would, in real life, have crushed its passengers to death, and their remains would have been incinerated by the heat generated as the craft sped up through earth's atmosphere. However, it is to Verne's credit that he at last partially foresaw some of these hazards. The first book ends with the firing of the giant cannon. Verne acknowledges the effects of high acceleration by making his travellers lose consciousness shortly after take-off. He incorporates a mechanism in the projectile which absorbs the recoil, using water as a damper – and because of this his astronauts survive.

In *Around the Moon* they recover consciousness and feel uncomfortably warm due to the air friction after blast-off.

Their leader warns them that they will soon suffer intense cold as they are already floating in space; then like many real astronauts after them, they thoroughly enjoy the weird experience of weightlessness.

The original intention had been to land the Columbiad on the moon, but, when they are some 4,500 miles from earth, a large meteor passes close enough to throw the capsule off course, and, instead of landing they find themselves in a moon orbit, with an excellent view of the moon surface, like Apollo 8. 'Powerful fireworks' inside twenty steel-lined guns are ignited, and instead of landing they accelerate back to earth, to achieve the first splashdown, in the Pacific – where they are recovered, with an incredibly fortuitous piece of foresight, by an American corvette.

Verne pictured the moon's surface as being as barren and inhospitable as it has now been shown to be, but he was mistaken about its atmosphere: he envisaged visitors to the moon casually wandering about, dressed for the country, and smoking cigarettes. After Verne's masterpieces, fictional accounts of moon trips declined in importance, and fell outside the mainstream of science fiction; the moon was beginning to be regarded as a dead world, and interest shifted to the planets. The first suggestion for an artificial manned satellite came only a few years after Verne's *From the Earth to the Moon*. In 1869 the *Atlantic Monthly* began to serialise Edward Hale's book, *The Brick Moon*. Two hundred feet in diameter and made of brick to withstand the heat, the brick moon was to be placed in polar orbit at a height of 4,000 miles. Here again fiction has become fact within a century, for this satellite was to serve as a navigational aid to sailors, permitting them to determine longitude with accuracy and ease. Hale proposed using a flywheel technique to hurl his satellite into orbit. After rolling down 'a gigantic groove' two flywheels would catch it simultaneously and fling it into space. Unfortunately the release was premature, and the brick moon carried several astonished workers and their wives into the sky. A year later people were seen jumping up and down on it, signalling their needs to the ground by Morse code, and Hale describes how their friends and relatives attempted to throw books and gifts up to them.

Ironically the next popular account of a moon journey is remarkable only for showing how far off the mark a celebrated prophet can be. H. G. Wells wrote *The First Man in the Moon* in 1901: the absent-minded Professor Cavor builds a sphere coated with a special gravity-screening substance, called Cavorite. By opening various blinds the sphere can be made to travel in any direction, and the Professor uses it to visit the moon. Even the Wellsian moon is far different from Verne's realistic picture, populated with insect creatures living in caves and tunnels, like Kepler's Subvolva.

At the time of Copernicus, astronomy was almost inextricably involved and confused with philosophy and pure fiction. Yet within the few centuries that separate him from

Jules Verne, the scientific method of using observation and experiment had become firmly established. That alone makes it all the more remarkable that the man who is regarded by many today as the father of rocket dynamics and the theory of space flight, should select Verne for credit as the major source of his inspiration. 'For a long time I thought of the rocket as everybody else did – just a means of diversion and of petty everyday uses. I do not remember what prompted me to make calculations of its motions. Probably the first seeds of the idea were sown by that great, fantastic author, Jules Verne – he directed my thought along certain channels, then came a desire, and after that, the work of the mind.' The writer of that admission was a Russian, Konstantin Tsiolkovsky.

Above *The faces of the Man in the Moon, as observed in the sixteenth century (a facsimile of a wood engraving attributed to Holbein).*

Right *Tycho Brahe at work in his observatory.*

EFFIGIES TYCHONIS BRAHE O.F.
ÆDIFICII ET INSTRUMENTORUM
ASTRONOMICORUM STRUCTORIS.
A°. DOMINI 1587, ÆTATIS SUÆ 40.

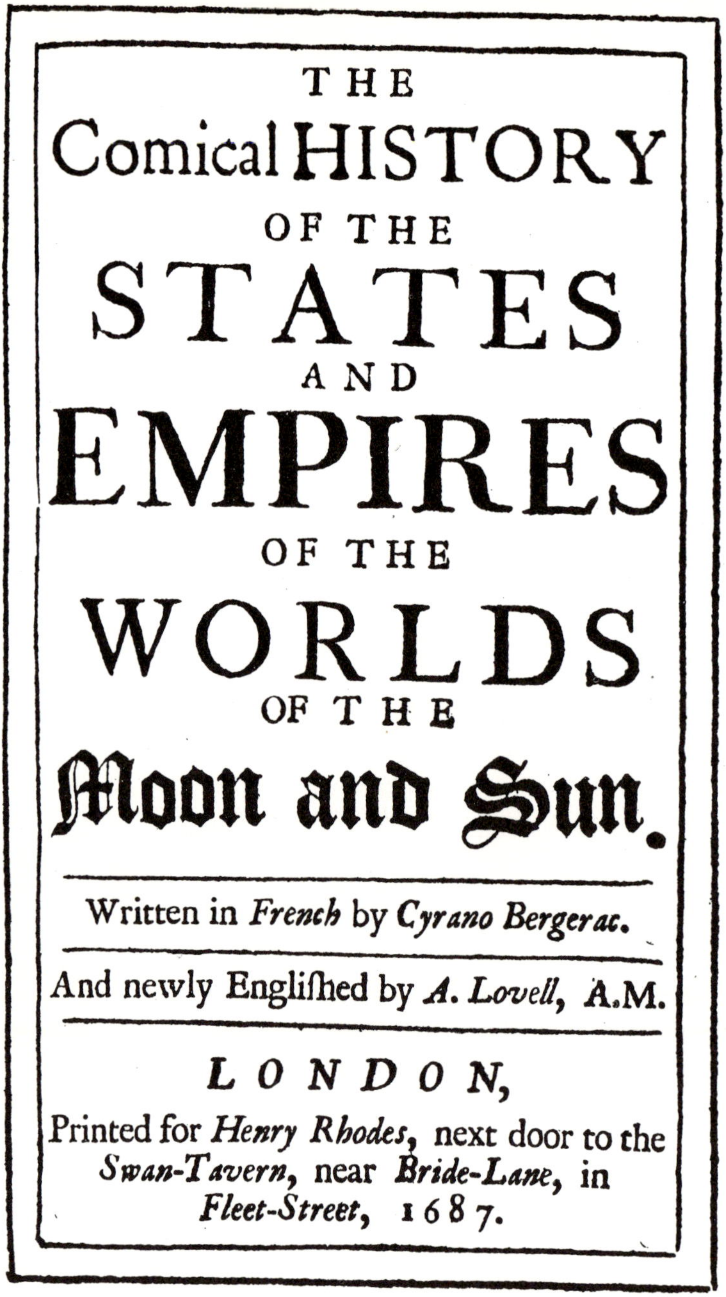

THE

Comical HISTORY

OF THE

STATES

AND

EMPIRES

OF THE

WORLDS

OF THE

Moon and Sun.

Written in *French* by *Cyrano Bergerac*.

And newly Engliſhed by *A. Lovell*, A.M.

LONDON,

Printed for *Henry Rhodes*, next door to the *Swan-Tavern*, near *Bride-Lane*, in *Fleet-Street*, 1687.

Title-page to Lovell's expurgated translation.

Above *The title page of Cyrano de Bergerac's* Comical History.

Right *The dew-bottle lift-off by Cyrano de Bergerac.*

Cyrano's first attempt.

The Flight to the Sun.

Left *De Bergerac's hot-air launch vehicle.*

Right *Domingo Gonzales in lunar transit, carried by a flock of geese.*

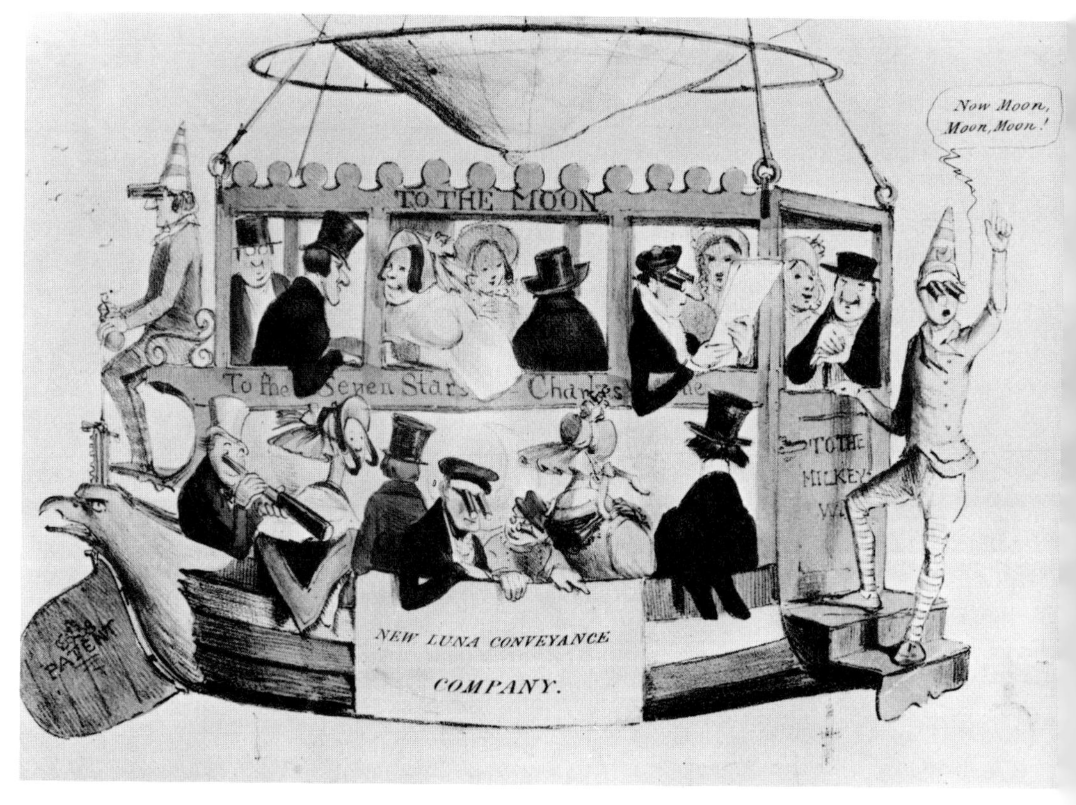

Above *A nineteenth-century prophecy of lunar travel, involving 'The New Omnibus Company, capital £12,000,000,000.'*

Above *Weightlessness as foreseen by Jules Verne.*

Above *Splashdown, another remarkable Verne prophecy of the 1870's.*

Konstantin Tsiolkovsky Russian father of space travel

The earth is the cradle of reason,
but one cannot live in the cradle for ever
Tsiolkovsky

Konstantin Tsiolkovsky was the first man to fully appreciate the significance of rockets and develop an understanding of them for use in space travel. Even in 1883, at the age of 26 he had understood the principles of reaction flight, and begun to realise its implications.

In many ways it is strange that it took so long for rockets to become the subject of serious scientific study: rockets had been employed for centuries for military purposes, and the reaction principle was in common use in fireworks. A Chinese legend of 2000 BC tells of a pair of kites, linked together and powered by forty-seven rockets, and it is certain that the Chinese were using fireworks by at least the first century AD. More than a thousand years later they used rockets when the town of Kai-Feng was besieged by the Mongols, and ten years after that in 1242, Roger Bacon invented an improved form of gun-powder to power the first medium range incendiary projectile. By 1815 the British army had its own Rocket Brigade, and it played a major role in the Battle of Waterloo. But it was not until the end of the nineteenth century that the true significance of rocketry began to be understood: in this achievement Tsiolkovsky has only two possible rivals for the honour of being first.

Nikolai Kibalchich took part in the assassination plot of 1881 against the Emperor Alexander II. He was arrested and sentenced to death: during his final days in jail he developed a scheme to propel a platform by rocket power. Gun-powder cartridges were fed into a combustion chamber attached to the platform, and on ignition generated an upward thrust. The direction of this motor was variable so that the platform could, theoretically, be propelled either upwards or sideways. It is not known what other research work he may have carried out, and he was hanged at the age of 27.

Hermann Gandswindt, a Prussian born in 1856, turned out invention after invention – he had no mathematical training, and many of his ideas were catastrophic failures, but somehow he evolved the notion that a reaction device could operate as well in space as on earth. He suggested a spaceship propelled by steel cartridges charged with dynamite – but failed to realise that combusting gases would have given sufficient power. Half of each cartridge would be ejected by the force of the explosion, leaving the other half to strike the top of the chamber and provide the motive force. The crew were to be housed in a chamber below, suspended by springs.

Tsiolkovsky realised the limitations of solid fuel for practical research. He proposed liquid fuels as an alternative, where one liquid becomes the fuel, and the other, the oxidant, enables the fuel to burn; expanding in a combustion chamber the reaction produces a hot exhaust gas which propels the vehicle. He even suggested the multi-stage rocket to overcome the weight limitations of carrying sufficient fuel for a long trajectory. He correctly foresaw the zero gravity situation of space flight, and for extended flights in spacecraft suggesting the use of green plants to remove excess carbon dioxide.

Born near Moscow in 1857, Tsiolkovsky became deaf at the age of nine, and withdrew into himself and the world of books which he found in his father's library. Even as a teenager he had become fascinated with the idea of interplanetary travel. At the age of sixteen he was sent to Moscow to further his education, and spent most of his allowance on books; he nearly starved himself to buy materials for constructing models, and chemicals for his experiments. In 1879 he settled down as a schoolteacher, and began to experiment in a home laboratory. Within four years he made the discovery which, if applied, could have led the world immediately towards space flight. He wrote, in 1883: 'Consider a barrel filled with highly compressed gas. If we open one of its taps the gas will escape through it in a continuous flow, the elasticity of the gas pushing its particles into space will also continuously push the barrel itself. The result will be a continuous change in the motion of the barrel. Given a sufficient number of taps (say six) we would be able to control the gas exhaust however we choose, and the barrel (or sphere) would describe a curved trajectory whose radius would depend on the barrel's velocity.'

In those lines Tsiolkovsky revealed a basic understanding of reaction flight; twenty years later his first article on rocketry appeared, entitled 'The Exploration of Space with Reactive Devices'. Unfortunately the article was so long that it had to take up two instalments of the *Scientific Review*, and before the second part could be published the magazine was seized as suspect of political overtones. The second part was eventually published in 1911.

He described several forms of rocket propulsion, but the first would be regarded as the most modern today – the propellant combination of part of the Saturn V, liquid oxygen and liquid hydrogen. In the second part he expounded his theory of a rocket's flight involving the change of the rocket's mass in the process of its movement. It was in this paper that he first discussed the possibility of rockets exploring the upper layers of our atmosphere. He suggested rockets for interplanetary communication, and proposed building and launching satellites to orbit the earth. Already in 1911 he was talking about jet propulsion and rocketry opening a new era in which man would be able to step on the soil of asteroids; to pick up a stone on the moon with his own hands; to organise interplanetary platforms orbiting the earth, the moon and the sun; to observe Mars at close quarters, and even land on it.

Later he discussed almost the whole arsenal of modern rocket fuels, suggesting not only hydrogen for use with liquid oxygen, but alcohol, methane and kerosene. He also suggested an alternative to the multi-stage rocket where each part is jettisoned once its fuel is fully burnt. He devised a system like a flying chandelier of linked rockets, all fired simultaneously; when half the total fuel is expended the outer rockets refill the inner ones, and drop off, the process being repeated until only one central passenger-carrying rocket is left, travelling at enormous speed.

Tsiolkovsky was clear that the future of space travel lay in some type of multi-stage, or chandelier, cluster of rockets: he calculated that for a rocket to reach what he called cosmic velocity (the term still used today by the Russians to describe the speed that must be attained for a vehicle to go into orbit, approximately 4·91 miles a second at ground level), the weight of fuel must exceed that of the rocket by at least four times. For a single-stage rocket to attain cosmic velocity this weight of fuel implies an immensely powerful motor, but if the weight of the rocket is constantly reduced due to the jettisoning of stages, a weaker motor can be equally successful.

But in Russia it was now the eve of the First World War, and officials paid little attention to his work. In 1914 he wrote: 'How difficult it is to work for years all alone under unfavourable circumstances, and not see any light or help from anywhere.'

However, in 1919, he began to gain recognition, and was elected to the Socialist Academy, and later granted a pension by the Soviet government. Between 1925 and 1932 he produced sixty works on astronautics, astronomy, mechanics, physics and philosophy. The ironic fact about Tsiolkovsky's life was that, although he spanned the entire period of the evolution of basic astronautical theory, he was not responsible for, and did not see, the major practical developments in liquid propellant rocket engines that took place in America and Germany in the 1920's and 1930's. There were many reasons for this, including Russian political and economic instability, lack of resources, and the failure of the military to appreciate the significance of his discoveries. As H. G. Wells said: 'For the scientific man at first the Soviet government had as little regard as the first French Revolution, which had "no need for chemists"'.

The curious aspect of the sudden development of rocketry was that it happened in several countries at almost the same time, and with the pioneers in virtually total ignorance of their rivals' parallel research work. Tsiolkovsky had perfected his multi-stage theories in detail by 1929: however, in America, Dr Robert Goddard had already worked out and made public one multi-stage principle fifteen years before, in 1914. Tsiolkovsky's reaction in 1929 is a poignant example of a brilliant mind discovering that his inventions were not original: 'I discovered much that had already been discovered before me. These discoveries are important to no one but myself, for they gave me confidence in my ability.

'For forty years I have been working on the reactive engine and I thought that a journey to Mars will begin in hundreds of years. But time perspectives change. I believe that many of you will witness the first trans-atmospheric journey.'

Above *The Russian Father of space flight, Konstantin Tsiolkovsky in the 1930's.*

Robert Goddard American pioneer of the first practical liquid-fuelled rocket

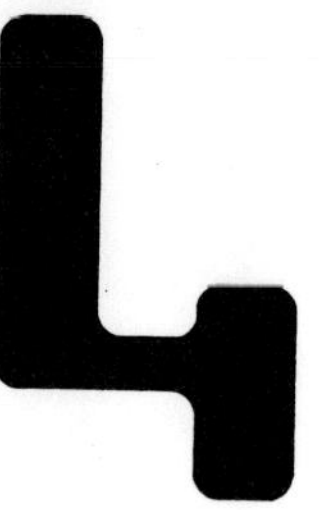

On 5 October 1882, a year before Tsiolkovsky described the principles of reaction flight, Robert H. Goddard was born. As a child he was prone to illness, and channelled his energy into reading, preferring the science subjects, and showing a precocious aptitude for mathematics and physics: it is significant that his favourite authors were H. G. Wells and Jules Verne. He even re-wrote Verne's *From the Earth to the Moon*, converting Verne's cannon into a rocket launch to send the astronauts into space.

At the age of sixteen he dismissed balloons as a means of high altitude transport after making a hydrogen-filled aluminium balloon which proved too heavy to fly. By the age of seventeen he was already trying to evolve a practical way of reaching the planets. Newton's inspiration is said to have come under an apple tree; Goddard had his pruning a cherry tree. He recorded: 'I was a different boy when I descended the ladder. Life now had a purpose for me.' By 1902 he had written an article entitled 'The Navigation of Space', and in another had suggested multi-stage spaceships as a means of space travel.

He continued his study of mathematics and physics, becoming increasingly aware that his early ideas were wrong; eventually he burnt all his models and notes, but could not lose his obsession with space flight. Just before graduation, an eruption of smoke from the basement of the Worcester Polytechnic Institute building signalled Goddard's first static test on a small rocket.

By 1909 at the age of twenty-seven he was studying for a Professorship in physics, and had already come to the conclusion that the ideal fuel combination for a rocket engine would be liquid hydrogen and liquid oxygen. He studied electrical theory during the day, and rocket propulsion at night. In 1914 he began fundamental research at Clark University, proving his theory that a rocket could function in a vacuum, and therefore operate in space; he was granted patents covering combustion chambers, nozzles, propellant feed systems, and multistage rockets. Such research was expensive and by September 1916 he had exhausted his private funds. Luckily the Smithsonian Institute in Washington happened to be interested at this time in high altitude devices for meteorological research, and having received a report on his experiments ('A method of reaching extreme altitudes'), asked for more information; Goddard provided it, and in return received a grant of $5,000 to continue his research.

When the United States entered the First World War, Goddard volunteered to direct his studies towards military purposes; on 7 November 1918, five days before the end of the war, he was at last ready to demonstrate several of his solid fuel missile prototypes at Aberdeen, Maryland. The largest was 5½ feet long, and weighed 50 lb. These missiles were later developed into the first effective hand-weapon to be used by infantry against tanks, the 'Bazooka' of the Second World War.

In 1920 a revised edition of 'A method of reaching extreme altitudes' was published, at the end of which a section referred to the space potential of rocket thrust: it contained a suggestion that a rocket similar in design to Goddard's could hit the moon. The proposal was concealed under the daunting title, 'Calculation of minimum mass to raise one pound to an "infinite" altitude'. Unfortunately for Goddard an enterprising journalist spotted it, and conjured it into a sensational story. The Press were amused, and many newspapers ridiculed Goddard and his work.

The adverse publicity had a deep and lasting effect, and all his future investigations were conducted in absolute privacy. Locked in a cabinet in a folder marked 'Formula for silvering mirrors,' he concealed papers like 'Speculation on the possibility of a rocket in space', 'Methods of navigating in space', 'The potential of the solar powered engine', and 'Manned and unmanned missions'.

By November 1923 he was ready to conduct his first static test of a truly practical, liquid-fuelled rocket engine. It was fairly successful, but he spent months modifying his apparatus, particularly the pumps. Next he had to devise and perfect methods for stabilising and guiding a rocket. Finally, on 16 March 1926 he was ready to test his invention in flight.

Accompanied by his wife Estha and two assistants he drove out to a farm near Auburn, Massachusetts, and prepared his rocket for launching. Ignition was effected by a blow torch, attached to a long pole. The rocket lit, and soared to a height of 184 feet, at a speed of 60 mph: it was the first flight of a liquid-fuel rocket in history – Goddard took great care to ensure the Press never heard of it. Goddard continued with his experiments, reporting to the Smithsonian Institute in 1928 that he had demonstrated the rocket's potential for the study of ultra-violet rays, and the composition of the upper atmosphere. Then, on 17 July 1929, Goddard launched another rocket: it rose 90 feet, and travelled 171 feet in trajectory, carrying a basic scientific payload of a camera, a barometer and a thermometer. The payload came down by parachute, but the rocket crashed in flames, and was observed by other people in the area who reported an aircraft crash. The result of this was that Goddard was forbidden by the State Fire Marshal from conducting any further experiments in Massachusetts. The Smithsonian Institute helped him by obtaining permission from the army for Goddard to experiment on federal property at Camp Devens, Mass. As a precaution against fire he was only allowed to launch rockets after rain or snow-fall.

By now Goddard had become increasingly angry with the Press hounding him for more stories about the 'moon-rocket-man's escapades'. However their colourful reports were a blessing in disguise: Goddard's research costs were rising, and up to now he had relied on the Smithsonian Institute for regular, but relatively small grants. The celebrated aviator Charles Lindbergh was interested in rocket power for aircraft, and read of the 'moon-rocket-man'. He visited Goddard and was so impressed with his work that he persuaded the Guggenheims to finance him. Through the Guggenheim Foundation, between 1929 and 1941, Goddard received about $150,000 for his research. The initial payment of $50,000 enabled him to move to New Mexico, to the Mescalero Ranch near Roswell, where he lived for the remaining fifteen years of his life, and was able to carry out his rocket tests with greater freedom. At Roswell he worked on gyroscopic rocket control and what he called 'Curtain Cooling', whereby the combustion chamber is cooled by the rocket fuel. But in spite of the new grants it was still research in a shoe string. One of his mechanics is reported as saying: 'There was never so much invention with so little manpower'.

In the Second World War he again offered his services to the army, but they were more interested in his Bazooka ideas, and his application of rockets for assisting aircraft during take-off: the army failed to appreciate the military potential of the long range rocket as a war-weapon, and when the first V-2's began to land on Britain they came as a cruel shock to the Allies. When the details of the V-2 reached Goddard he noticed similarities between the German weapon, and his own liquid-fuel rocket: in basic design they were almost identical. Few people were prepared to quibble about patents in war-time, but ever since the war there has been a controversy over just how much the Germans made of Goddard's patent designs. He died on 10 August 1945, and ironically the Press did not even notice the demise of America's 'moon-rocket-man' although he did as much to start the Space Age as the Wright Brothers did for aviation. He registered more than 200 patents, many of which were issued after his death, and cover the whole field of rocketry. In 1960 the United States government somewhat belatedly gave Goddard the credit that was due to him by granting his widow and the Guggenheim Foundation a million dollars for the use of his patents. Very appropriately however, on 1 May 1959, when the National Aeronautics and Space Administration named its new Space Flight Centre at Greenbelt, Maryland, they called it the Goddard Space Flight Centre.

5

The V-2 Story Herman Oberth and Wernher von Braun

Left *16 March 1926.*
Dr Robert Goddard ignites the first successful liquid fuel rocket at Auburn, Mass.

In 1905 an eleven-year-old boy was given by his mother Jules Verne's books *From the Earth to the Moon*, and *Round the Moon*. He read them until he knew them by heart. His name was Herman Oberth, the third great pioneer of rocketry and astronautics.

Shortly before the First World War he became interested in the use of rockets for military purposes, and in 1917 he proposed a long range, liquid-fuelled missile to the German War Department. In 1922 he wrote to Goddard requesting a copy of his 1919 publication – a request which was to worry him in later life in case he was suspected of copying Goddard's ideas. By 1923 he had already published his first book, called *The Rocket into Planetary Space*. He studied problems ranging from space navigation to guidance and landing techniques; he drafted the designs for a two-stage unmanned rocket, called the Model B, and even conceived a two-passenger spaceship with an egg-shaped cabin in its nose which would return the spacemen by parachute. The stages of this rocket were to be powered by oxygen/alcohol, and oxygen/hydrogen. Although it was only a 92-page book it contains many brave assertions which we accept today as proven fact. In 1929 an expanded edition of 425 pages, called *The Road to Space Travel*, was published: these two volumes were to become highly significant in the growth of interest in space flight for they inspired many of his contemporaries to pursue similar lines of research.

However, even by 1927 interest in rocketry had grown sufficiently in Germany for the VFR (Verein für Raumschiffahrt: Society for Space Travel) to be formed: it was a society devoted to the study and pursuit of space travel, and its President, Johannes Winkler, later became the first man in Europe to fly a liquid-propelled rocket. Soon there were 500 members, and the society published a journal called *The Rocket* to raise money for converting some of its theoretical projects into tangible facts. In 1930 the American Interplanetary Society was founded in New York, and by 1933 its British counterpart was also formed – with George Bernard Shaw as one of its more improbable members.

Rocketry and science fiction were the novelties of the age, and in 1928 the film director, Fritz Lang, decided to make a film called *Die Frau im Mond* (The Woman in the Moon). The UFA film company approached Oberth, asking him to design, build, and fly a rocket. The actual rocket launch was to be the climax of the publicity campaign coinciding with the première of the film on 15 October 1929. Oberth started work with two assistants in the autumn of the previous year; he may have been a great theorist, but he was certainly no great engineer, and neither were his two assistants. The rocket was eventually built, but it became clear, due to the continuous modification of its components, that it would never fly before the première. In spite of this however, the film was an enormous success, and contributed greatly to the ever growing interest in rocketry.

Max Valier, who had originally suggested the formation of the VFR had succeeded in interesting the industrialist Fritz von Opel in rocketry, and in May 1929 Opel himself drove a car powered by 24 powder rockets at a speed of 125 mph. Valier persisted for a while with powder rockets as a means of propulsion, but by the end of 1929 had begun to experiment with combustion chambers for liquid fuel. While studying the propellant power of a mixture of paraffin and liquid oxygen in his laboratory, Valier was killed by an explosion on 17 May 1930. A very similar fuel today powers the gigantic first stage of Saturn V.

By now Winkler was pioneering the use of methane with liquid oxygen, and on 31 March 1931 his first liquid-fuelled rocket rose to a height of 1,000 feet. Experts in Europe were at this time quite unaware that Goddard had already launched his liquid-fuelled rocket five years before.

At this time a seventeen-year-old engineering student called Wernher von Braun volunteered to help Oberth move his test equipment from the film studios to a disused ammunition dump near Berlin which was to become known as the Raketenflugplatz (literally the Rocket Flight Place): it was to become the site of the experiments which eventually led Russia and the United States into the conquest of outer space.

In August 1930 tests began on the tiny Mirak rocket. The body was a foot long with a three-foot tail holding the petrol tank. The head stored the oxidiser (liquid oxygen) which also cooled the tiny copper combustion chamber where the fuels would mix and explode. By 1931 the rocket had failed on two static tests after the combustion chamber had burnt through.

A new series called Repulsor was begun: its motor was fitted with aluminium walls and cooled by water instead of liquid oxygen. On 14 May 1931 Repulsor 1 reached a height of 200 feet; Repulsor 2 was slightly more successful, and Repulsor 3 climbed to over 2,000 feet.

By the summer the design had been improved by adding a single stick like a Guy Fawkes' rocket, for greater stability, and on 31 August Repulsor 4 rose to 3,300 feet and parachuted safely back to the ground.

Even in the early pioneer days, rocketry was an expensive hobby and with the Great Depression the VFR began to find it difficult to finance research; also the Berlin police were making increasing objections to the rocket flights within the city limits. The VFR membership dropped alarmingly and the only source of support seemed to be the German Army, and in the summer of 1932 a Repulsor rocket was demonstrated at their nearby proving grounds at Kummersdorf.

The Army had good reason to be interested: under the Treaty of Versailles Germany was forbidden to manufacture heavy artillery. Professor General Karl Becker was Chief of the Army's Weapon office, and had written a book on ballistics, devoting one chapter to the use of rockets as long range weapons. Becker had already started a small Army department to study rocketry progress; after the Kummersdorf demonstration von Braun was invited to join the Army and continue his thesis on rocketry.

Hitler's rise to power and the attentions of the Gestapo meant the virtual end of the internationally oriented VFR, and in January 1934 the Raketenflugplatz returned to being an ammunition dump.

The first Kummersdorf rocket, the A-1, never flew. After a few static tests the liquid oxygen/alcohol engine blew up. The A-2, stabilised by a 70-lb. gyroscopic flywheel, rose to over 6,000 feet. General von Fritsch, on a visit to Kummersdorf, was impressed with the work. The unit was granted more funds and transferred to a quieter, secret site, away from Berlin. Then a new research establishment was set up in 1937 on an island in the Baltic. The base was given the name of a nearby fishing village, a name which was soon to become synonymous with the most sinister aspects of rocketry: Peenemunde.

Von Braun became the technical director of the Army Experimental Station there. He immediately concentrated on building a 3,300-lb. thrust alcohol/liquid oxygen rocket, the A-3. All test prototypes were failures, although the news did not leak out for many years. The Peenemunde project had now been classified as top secret.

In March of 1939 Hitler visited the rocket research station at Kummersdorf-West but was unimpressed and did nothing to accelerate the construction or testing of rockets. High altitude rocketry was low on the war-time priority list, although the Army Ordnance Department suggested they develop a missile capable of flying 200 miles with a one ton warhead; the size of the rocket had to be compatible with rail transport to distant launch pads, and it had to be completely reliable. The resulting design, the A-4, was later to develop into the notorious V-2 of the Second World War.

The failures of the A-3 guidance system persuaded von Braun to perfect a further series of test vehicles, the A-5's, before continuing with the A-4 project. From 1938, while the large motors for the A-4 were being designed and constructed, trials were carried out launching the smaller A-5 rockets from Heinkel bombers and from the ground. In the summer of 1939 the A-5 made its first vertical flight, reaching a height of over seven miles. Other test models performed equally well – some were launched vertically, some on a slanting trajectory to simulate the A-4, and some were even recovered by parachute and used again. By 1940 at least 25 A-5's had flown, and many components of the A-4 were already in production.

In 1942 the first V-2 emerged from the Peenemunde assembly plant. It was the largest, most advanced rocket in the world, weighing over 27,000 lb., and standing 46 feet high. Powered by liquid oxygen and alcohol it could carry a 2,200-lb. payload some 210 miles.

After two failures the third launch on 3 October 1942 was a complete success – the A-4, as it was still called, rose over 50 miles, and crashed 120 miles away. The commander of

Peenemunde, Walter Dornberger, waited for one more successful flight, and then went to the Minister of Munitions, Albert Speer, to try to persuade him to give the rocket research establishment more assistance. Fortunately for the later survival of London, his entreaties got him almost nowhere, and an empire-building committee of bureaucrats was set up which nearly buried Peenemunde in red tape. The situation was aggravated by Hitler having a bad dream about rockets, and cutting the supplies to the research establishment.

Finally in May 1943 Speer, Admiral Dönitz and Field Marshal Milch witnessed comparative tests between the 'doodle bug' flying bomb (the V-1) and the A-4. Although the flying bombs were complete failures on that occasion contracts were signed for the mass production of both types of weapon, and Hitler decided to forget his nightmare. However, before huge numbers of the V-2 could be made, the RAF raided Peenemunde, dropping 12,000 tons of bombs on the rocket station, killing 800 members of the staff including the director of engine development, and badly damaging the assembly plant. V-2 production began again in a converted oil depot deep under the Harz Mountains near Nordhausen. By the end of the war 900 V-2's were being turned out every month, but thanks to the raid on Peenemunde the first V-2's did not land in England until 8 September 1944. They were first launched from a site near the Hague in the Netherlands at the rate of two a day.

Altogether about 3,000 V-2's were built; more than 1,500 dropped in Southern England killing 2,500 people and causing immense damage. If all the systems functioned correctly the V-2 landed on its target some five minutes after launching at a speed of 3,500 mph. The last one fell on Orpington, Kent, on 27 March 1945, but by then the Germans could no longer afford their rocket programme: it has been estimated that the first V-2 cost £100,000,000 to develop, and its design needed 60,000 modifications before it could be flown in combat.

German defeat seemed certain by January 1945, and the staff of Peenemunde had the choice of awaiting the arrival of the Russians or going south to surrender to the Americans. Von Braun and the majority of his colleagues chose to go south. Amid a deluge of conflicting orders they bluffed their way through the Gestapo check-points to reach Bleicherode in the Harz Mountains where it had been arranged that rocket research should continue. The death of Hitler on 30 April finally persuaded von Braun and Dornberger to surrender, and months of interrogation began for the captured scientists.

American, British, and Russian intelligence were all very anxious to find the German rocket men, and capture as many intact rockets as possible. The Americans called the project Operation Paperclip and were astonished to enter the huge underground assembly plant at Nordhausen and find its lines of V-1's and V-2's completely undisturbed. In military terms it was like the discovery of Tutankhamen's treasure. Arrangements were made to remove as much as possible before the arrival of the Russians, and a hundred V-2's were ferried by train to Antwerp to be loaded aboard sixteen liberty ships bound for New Orleans.

The Red Army succeeded in 'recruiting' about 5,500 of the remaining personnel at Peenemunde, and set up a temporary research establishment at Bleicherode before moving the entire organisation back to Russia. But, in all probability, it was the United States that gained most from the enforced demobilisation of the German Rocket Establishments, since Dornberger and von Braun took with them to America their plans for the continuation of the 'A' series. However, the ignorance of American troops at Nordhausen meant that the Russians were able to capture several priceless blueprints for very advanced rockets, together with some unique machine tools.

By the autumn of 1946 Stalin realised that many key German scientists had slipped through the Russian net. The secret police under General Serov brought back 6,000 German specialists in nuclear science and rocketry to Russia within the year. It was to be several years before the West saw the results of their work in collaboration with their new Russian colleagues.

In January 1946 the US Army announced a firing programme of its newly acquired V-2 rockets at White Sands, New Mexico. Several universities were invited to use the V-2's for high altitude research. The rockets enabled soundings to be made to a height of 100 miles, measuring the high-energy particle radiation which is found at such altitudes. Within 6 years over 60 were launched and the United States had learnt how to make similar rockets. By 1947 the Naval Research Laboratory in Washington DC had designed a smaller, more powerful launch vehicle, later to be known as the Viking; it flew to an altitude of 158 miles, carrying a larger payload than the V-2. However military experts regarded the rocket as having little significant potential. It seemed impossible to construct a rocket capable of lifting the heavy A bomb – and 'no explosive force short of an atomic one would inflict enough damage to warrant the expense of putting it into orbit'. Because orbiting satellites did not appear to be potential weapons, there were no funds available for rocket development. None of the three services was allowed to continue development of a rocket capable of putting satellites into orbit, and satellite studies were ignored at the Department of Defence. In November 1954 its Secretary remarked publicly that he knew of no American satellite programme.

In 1954 it was proposed that the International Geophysical Year should take place in 1957–58. An artificial earth satellite was suggested and both the United States and USSR undertook to launch one. In September 1955 the Navy was given authorisation for Project Vanguard. The first stage was an adapted Viking rocket, carrying another rocket which owed much to the V-2, an improved Aerobee, with a solid fuel third stage. Two and a half years later, on 17 March 1958, the first successful Vanguard satellite was launched. But by then the Americans had found just how far Russian Space technology had advanced. Five months before Vanguard lifted off, Sputnik 1 was in orbit.

Postwar Secrecy: Sputnik and Gagarin. Russian research forges ahead. The US loses face.

Russian rocketry was never just a series of developments which took place in the wake of the German V-2 projects of the Second World War: Tsiolkovsky is occasionally accepted in the West as having been a genuine pioneer, but his work is often regarded as being outside the mainstream of significant research. This is largely because the remarkable growth of interest in rocketry in the Russia of the 1920's was generally unknown in the West until recently. Many Western experts were astonished to find that not only were the Russians working on liquid-fuel propulsion but that they were experimenting with highly sophisticated electric arc rockets, a type of propulsion which is receiving much attention today as being perhaps the next stage of power for space travel.

In 1929 the Gas Dynamics Laboratory was set up in Leningrad to work on electric propulsion theory, and the project was only dropped when it became apparent that such vehicles could never become efficient until they were in space. Some other type of rocket would be needed to launch them above the atmosphere. In August 1933 a group was formed for research into rocket motion. It was called GIRD, and launched the first Soviet liquid-fuelled rocket: powered by burning petrol in liquid oxygen, it rose to a height of more than 1,200 feet. The head of the launch team was a scientist called Sergei Korolyev, later to receive a State Funeral in 1966 in honour of being the chief designer of the rockets that put Sputnik 1 and Yuri Gagarin into space: Party chief Leonid Brezhnev and President Nikolai Podgorny helped carry his coffin to rest within a few yards of the grave of Stalin. His name was never mentioned in the Press, and foreign diplomats were mystified when thousands of Russians braved a snowstorm in Red Square to witness his funeral. If any one man could be credited with being the brains behind Russia's conquest of space, that man would almost certainly be Sergei Korolyev. By 1933 he was head of the newly formed State Reaction Research Institute, although his early interests were mainly in rockets as a means of powering gliders. Tsiolkovsky's ideas had fascinated him, and by 1934 he wrote a thesis on high altitude flight, suggesting the development of really large liquid-fuelled rockets: he also predicted that acceleration, and pressurisation might present fundamental problems for manned space flight. Just before the Second World War it is known that he was employed on designing and perfecting a Russian flying bomb, but at this high point of Stalinist rule little is known about the exact details of his work.

In October 1946 the remnants of the German V-2 manufacture and design team were rounded up and taken back to Russia. Once there it was hoped they would help the Russians radically improve the performance of the V-2. Wholly Russian design teams were working at this time on the R-14, the Pobeda (Victory) Rocket: eventually it proved so successful that (apart from mass copying of the V-2's) most of the other V-2 work, involving improvements, was dropped. However most of the Germans were detained in Russia until 1953 before being allowed to return home.

It is difficult to evaluate the importance of German help in post-war Russian rocketry, but it is intriguing to note that the German team designed a modified Pobeda rocket of a conical shape: superficially it was very similar to the design of each individual rocket in the cluster that was later to make up the Vostok 1 launch vehicle.

Stalin was not unaware of the military significance of the V-2, but wanted its range increased, and research was begun in Moscow on the intercontinental ballistic missile: again Sergei Korolyev became involved as one of the project design chiefs. The almost impenetrable secrecy of the Iron Curtain now made it virtually impossible for the West to realise how energetic the Russian programme of rocket development was becoming by the end of the 1940's: only in 1967, when the Vostok launcher was shown in Paris at the Le Bourget Air Show, could experts appreciate just how advanced the Sputnik launch vehicle was. Each of its 20 main thrust chambers was four times as efficient as the old V-2 in producing high combustion pressure. The Russians had evolved a system of standardising their engines, giving them the great advantage of being able to mass-produce well proved components which could be assembled to make up a wide variety of rocket engines for both single- and multi-stage vehicles.

However, at the end of the Second World War virtually no one in the West could foresee the Russian appetite for knowledge and experience of advanced rocketry. Germany was a defeated nation, and the sinister aspects of Communism could continue to be overlooked for many years whilst Russia recovered from the ravages of a war which had forced her into an expedient if improbable alliance with the West.

Yet, even in 1945, two organisations were already seriously studying the possibility of putting unmanned satellites into orbit. The first was the United States' Navy; it commissioned the Guggenheim Aero Laboratory at California Institute of Technology to research the idea. Results showed the proposition to be theoretically feasible, but far too expensive. The Navy therefore suggested a cost-sharing project to the US Army Air Corps; after considering the idea for a month the Army Air Corps vetoed it. It was some time before the Navy found out that the Army Air Corps had been the other organisation in question: it had already begun its own studies of artificial satellites. In 1946 a War Department Board heard of this duplication of effort, and made no effort to coordinate the work: it seems astounding today, but both the Army and the Navy were given full permission to carry on with their own private research programmes, and no pooling of ideas was enforced.

The Army decided to work on multi-stage rockets, and the Army Air Corps gained its independence from the Army to become the US Air Force. The Navy were still trying to get enough money to build a single-stage vehicle, which would orbit as a satellite once its fuel was spent. In 1947 a Committee on Guided Missiles was appointed to try to bring the services together into a coordinated plan, but even that bid at ending the time-wasting rivalry of the services was finally thwarted, when, in 1948, it was inferred by a Chief of Staff that the time was not yet ripe for a satellite programme. Theoretical discussions continued endlessly, almost always ending in a stalemate – a situation aggravated by the fact that the United States possessed no vehicle capable of launching a satellite, even if a policy could have finally been hammered out. Bad intelligence work was one major reason for this extraordinarily dilatory attitude: Russia was still regarded with benign complacency as the poor ally of the war, too socially confused and poor in both finance and intellect, to embark on any sophisticated policy of serious scientific or military significance.

Then during the early summer of 1953 intelligence reports and rumours were anxiously received at the Pentagon that the Russians were developing a long range ballistic missile, and that the programme was far ahead of anything comparable in the United States. Finally the United States Department of Defence was given an even worse shock: on 12 August 1953 the USSR exploded a hydrogen bomb. The most expensive peace-time arms race in history was accelerated.

The death of Stalin also meant a slight lift in the curtain of secrecy, and some Western experts must have wondered what significance to attach to a statement by the President of the Academy of Sciences on 27 November 1953 Professor Nesmeyanof said: 'Science has reached a point where it is realistic for us to speak of sending a stratoplane to the moon and of creating an artificial satellite for the earth.'

Each of the three US services began to push ahead with a missile programme designed to meet its own individual requirements: the Navy wanted missiles to fire from ships; the Army concentrated on short and intermediate range weapons, and the Air Force on long range – like the Atlas Intercontinental Ballistic Missile (ICBM) as it became known in 1954.

The Army and Navy joined forces, undertaking a joint development which eventually produced the Jupiter series. By September 1955, benefiting from the technological experience gained in pioneering the Redstone rocket, the first firings of Jupiter A took place: it was an Intermediate Range Ballistic Missile (IRBM) capable of altitudes of 300 miles, with a trajectory of 1,200 miles. Jupiter C included an elongated Redstone making up the first stage; a second stage was a cluster of 11 solid fuel rockets which were placed around the third stage, 3 solid fuel rockets that spun in flight for stability. However the 110,000-lb., 58-feet long rocket seemed highly unsuitable for launching from a submarine. In September 1956 the Atomic Energy Commission had announced a breakthrough in nuclear technology, prophesying that by about 1964 much smaller warheads would be available; because of this the Navy withdrew from Jupiter to develop Polaris, a smaller, solid propellant missile suitable for submarine firings.

Considering the administrative changes on the Jupiter series it is surprising it ever took off at all, for, in November

1956 the Secretary of Defence transferred the operational control and development funds to the Air Force, leaving the Army to work on surface-to-surface missiles with ranges of less than 200 miles. However by May 1957 the third Jupiter had succeeded in flying 1,600 miles. The Air Force now had two major ballistic missile programmes – but had proved its competence a year and a half before, when its third attempt at launching the Atlas ICBM went without a hitch.

However, while the United States were now less troubled by the need for monster launch vehicles because of their relatively light hydrogen bomb, the Russian lack of such a warhead forced them to develop the colossally powerful rockets which were later to be instrumental in giving them a commanding lead in the space race.

On 29 December 1955 President Eisenhower finally approved plans to launch an artificial satellite for the International Geophysical Year. As the Air Force Atlas was still unproved, von Braun, under the Army, suggested 'Project Orbiter', which would have made use of his own development, the Jupiter (IRBM). But the Navy won the day with Vanguard, and the United States confidently began to prepare for the first launch. Carrying dummy upper stages, the first Vanguard carrier was launched on 23 October 1957: it lifted a load of 4,000 lb. to a height of 109 miles – not exactly a milestone in the history of rocketry, but certainly a promising start to the series. However it was regarded as a catastrophe for American prestige. Eighteen days earlier a 23-inch sphere had started to spin round the earth once every 96 minutes. The Space Age had begun.

The London *Times* heralded the event with characteristic sobriety. The headlines read **'Russia launches earth satellite; 180 lb sphere travelling at height of 500 miles'**. Most newspapers were caught unawares and simply stated the bald facts of **'How to see it'** – the surge of comment began next day when the Sunday papers interpreted the significance of Sputnik 1. *The Sunday Times* achieved a classic understatement with **'Soviet lead in rockets says US'**. In typical English fashion the *Sunday Dispatch* managed to convert a Russian achievement into a British failure – **'Our scientists were caught napping by satellite'**. The article went on, 'British scientists are angry because Russia launches her man-made moon without warning. They regard it as the biggest "double-cross" of the century.' The *Dispatch* also noted the feeling in the United States which was to become one of the major elements of the later race to the moon. 'In America, Senator Symington, of Missouri, called for a full-scale congressional investigation into America's failure to win the space-race. "This cannot be laughed off", he said. "Unless our defence policies are promptly changed the Soviets will move from superiority to supremacy. If that ever happens our position will become impossible."'

The Observer noted the Cold War aspects of the flight – **'US eyes on military implications'**. It reported: 'Publicly, both scientists and military experts in America are tending to play down the significance of the Soviet satellite. But privately, western scientists attending the International Geophysical Year conference on rockets and satellites in Washington are completely astonished by the Russian announcement. Of immediate concern here (New York) are the military implications. Experts have already stated that the satellites would have no such practical implications. But the public is disturbed. It is likely to affect radically the defence cutting programme.' In England the giant radio telescope at Jodrell Bank was not yet complete and therefore could not follow the flight: the director of the research station, Professor Sir Bernard Lovell, was quoted as admitting: 'It is very frustrating,' although he acknowledged Sputnik 1 to be 'a brilliant scientific achievement'.

By Monday, 7 October the Press were hunting for political reactions to the news. One report regretted that President Eisenhower had 'not yet commented directly on the Russian success'. *The Daily Telegraph* went further: **'President is not alarmed: week-end golf** President Eisenhower has taken Russia's success . . . calmly. Like Drake when told of the Armada's approach he did not allow the news to interrupt his game (golf, not bowls).' The report added, 'American plans for launching its own satellite would not be changed nor speeded up. It was wrong to take the view that Russia and the United States had been engaged in a race which the United States had lost. The official calm is not reflected in political and some scientific comment. The launching of the satellite meant, some said, that Russia was not bluffing when she announced that she had perfected an intercontinental missile.'

Since 1956 von Braun and his team had been pressing to be allowed to expand the Jupiter C (a three-stage rocket) into a four-stage vehicle. He had been continually refused permission to develop any vehicle designed for space. Now, three weeks after Sputnik, and two days after the Vanguard first-stage launch, the Defence Department changed its mind. Von Braun was given the go-ahead: his team produced Juno 1, a rocket with a liquid propellant first stage, plus three solid fuel stages above. But before Juno or Vanguard could put a satellite into orbit, the Russians had proved that they were even further ahead than anyone had imagined possible.

On 3 November 1957 Sputnik 2 was launched into orbit, carrying not just a radio transmitter, but the first animal in space, a dog called Laika. The capsule revealed the extent of Russia's rocket power, Sputnik 2 was about six times as heavy as Sputnik 1, weighing over 1,100 lb. Laika's death on the sixth day was reportedly due to overheating of the experimental capsule through a faulty valve temperature control. After orbiting the earth for 162 days the capsule was finally destroyed when it burnt up on re-entry.

Only a few days before the launch of Sputnik 2 the go-ahead was given for Juno 1: the American 'loss of face', after Russia's two astonishing displays of technological prowess, was very considerable. Drastic measures were taken to launch a satellite: by 30 January 1958 Juno 1 had been made ready to carry Explorer 1. Bad weather postponed the flight, but the next day the United States was able to enter the lists, as Juno 1 left its launch pad and placed America's first satellite into orbit. The Navy Vanguard team were almost ready to launch their satellite, and after years of preparation their frustration must have been enormous when von Braun, a late starter, beat them on the final straight. But the Army's Explorer 1 weighed only 18·2 lb., compared with the 1,100-lb. payload of Sputnik 2. However from a scientific point of view Explorer 1 was highly important: it carried instruments designed by Doctor Van Allen to study cosmic radiation within a few hundred miles of the earth. Explorer 1 discovered that the earth was surrounded by belts of intense radiation, and many experts feared for some time that they would form an insuperable barrier for manned space flight to other parts of our solar system. The radiation zone was named after Dr Van Allen, and called the Van Allen Belt.

The era of satellite exploration had begun. Vanguard left the ground six weeks after Juno, putting the fourth satellite of space history into orbit on 17 March 1958. It is still in near space today, reaching over 2,500 miles from earth at the highest point in its orbit: using the sun's rays as a source of energy to regenerate its solar cells, the tiny 3-lb. satellite transmitted cosmic data back to earth through a minute beacon transmitter. Nine days later the Russians ended their 1958 satellite programme with Sputnik 3, an enormous 2,900-lb. orbiting laboratory, which relayed valuable information back to earth for almost two years; then at the very beginning of 1959 Russian scientists announced their intention of sending a spacecraft past the moon; on 2 January Moscow radio proclaimed the launch of Lunik 1. It was the first man-made vehicle ever to free itself from earth gravity. It is still in space today, revolving the sun once every 450 days. It missed the moon by over 4,000 miles but enabled scientists back on earth to do some valuable research into the precise density of outer space: it created the first artificial comet, by releasing a cloud of sodium which could be observed and statistically analysed through special filters.

'I am carrying out observations of the earth. Visibility is good, reception excellent. The flight continues well.' Those words had an even greater effect on the West than the 'bleep bleep' of Sputnik 1; they were spoken by Yuri Gagarin, from space. With the successful flight of Vostok 1, a single orbit of the earth lasting 108 minutes on 12 April 1961, Russia began the era of manned space flight.

Gagarin was launched in a spherical capsule, 7½ feet in diameter, aboard the multi-stage Soviet Vostok rocket, from Baikonur in Southern Siberia at 9.07 a.m. Moscow time. Forty-five minutes later he reported that he was over Cape Horn, and by 10.15 he had reached South Africa. At 10.25 the retro rockets were fired to put the spacecraft into a re-entry trajectory, and by 10.35 the capsule was plunging back into the denser layers of the atmosphere to land at 10.55 in the Sarato region of the USSR.

The year before, Sputnik 5 had carried the dogs Strelka and Belka on a day's outing in space, watched by on-board TV cameras, but even that momentous flight had not prepared the world to expect a manned mission quite so soon. It was later revealed that five experimental launchings were made before Gagarin's flight, and not all of them were successful: on 15 May 1960 after a Sputnik containing a dummy astronaut had entered orbit, a retro-rocket fired accidentally and sent the capsule into a higher orbit instead of initiating the re-entry trajectory.

Because of the Russian tendency to announce their achievements only after the event, there was a degree of carping criticism in the West. On 19 April 1961 the *New York Herald Tribune* published a headline **'Did Gagarin Do It?'** The article read: 'Was the Soviet stunt in outer space, as announced officially from Moscow, a hoax? Granted that something went around the earth, was a man really in it, or did the astronaut merely make a separate flight similar to that which an American airman, Joseph Albert Walker, recently made in an X-15 rocket plane at an altitude of 32 miles?

'These questions are being asked by scientists because there are some obvious discrepancies in the boastful account of his trip given by Major Gagarin. . . .'

Two British newspapers, *The Times* and *The Guardian*, even rumoured that there had been previous, unsuccessful manned flights: under the banner **'Major Gagarin "Not First"'**. *The Guardian* reported: 'Rumours that Major Gagarin's space flight had been preceded by an unsuccessful attempt received further currency yesterday in a French broadcast by Mr Bobrovsky, a radio correspondent who has just returned from Moscow. Mr Bobrovsky reported that his Moscow sources, which were "entirely reliable", had said that the Soviet test pilot Sergei Ilyushin, son of the aircraft designer, was sent into space three or four days before Yuri Gagarin. He came back to earth after three circuits, "having completely lost his sense of balance", and was now lying unconscious in a Moscow hospital.' These rumours were never disproved, and succeeded only in tarnishing the outstanding triumph of Gagarin's flight.

After his landing and post-flight checks Gagarin was rushed back to Moscow for a hero's welcome, including much hugging and kissing by a jubilant Mr Kruschev. Then began an exhausting, protracted world tour, with rapturous crowds in virtually every country he visited. The cosmonaut probably did more for Russian goodwill and prestige on that occasion than any man before or since. On his visit to London he showed that he was far from being a mindless guinea-pig, and at a long Press conference he showed an adroit ability at coping with awkward questions: at that time he was still able to say, about the problems of becoming a celebrity overnight, 'I am still an ordinary mortal, and have not changed in any way'. He made the flight at the age of 27, but by the time he died in an aircraft crash at the age of 34, photographs of him showed that the sudden limelight, banquets and honours had aged him considerably.

Four months after Gagarin's historic flight his statement that 'the completion of the flight opens new perspectives in the conquering of the cosmos' was proved true by the flight of Vostok 2. On 6 August 1961 Major Gherman Titov orbited the earth for 25 hours 18 minutes, circling the world 17 times and travelling nearly half a million miles. In terms of the space race it suddenly began to look as if the United States was a non-starter.

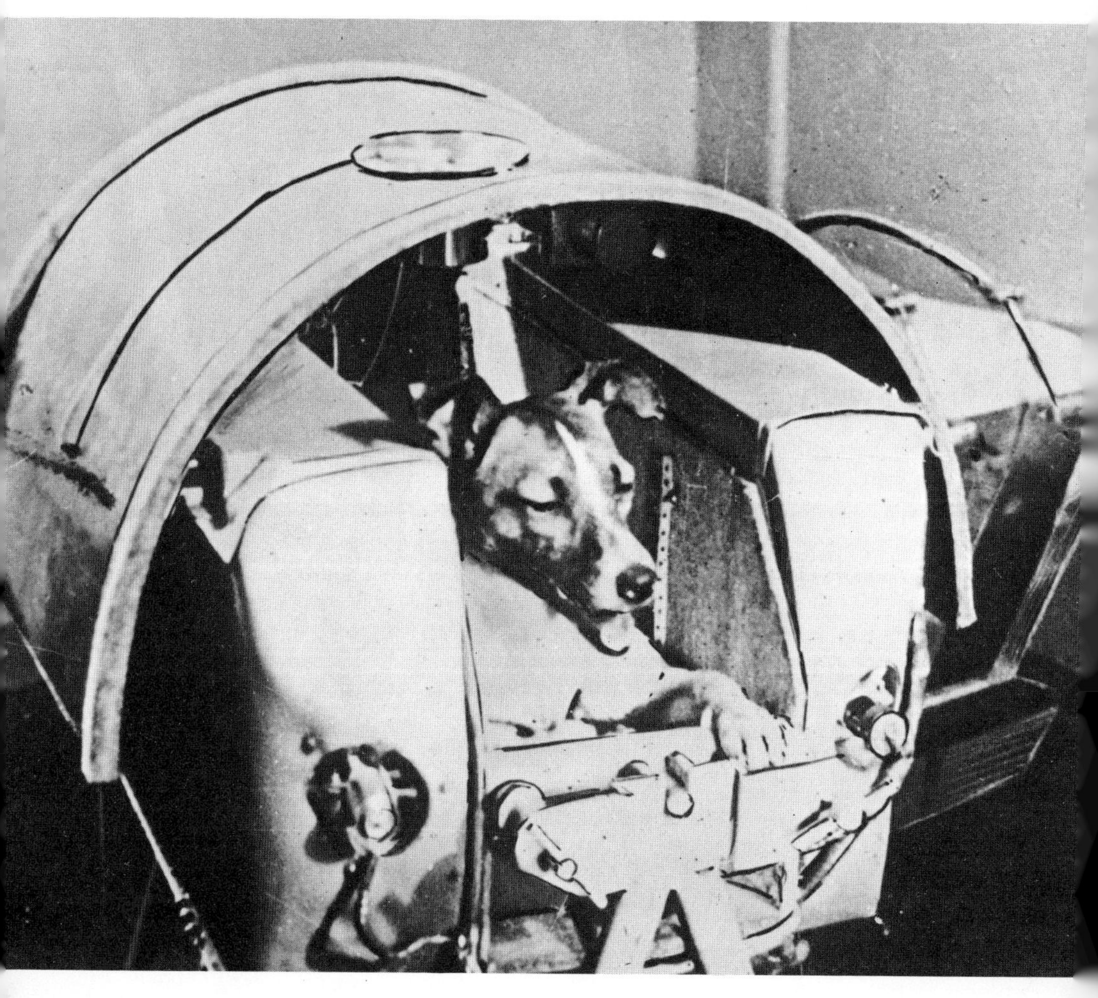

Above *Laika aboard an airtight compartment before being installed on the second Soviet artificial earth satellite (3 November 1957).*

Above *Strelka and Belka, four-legged cosmonauts of the second Soviet spacecraft to contain animals.*

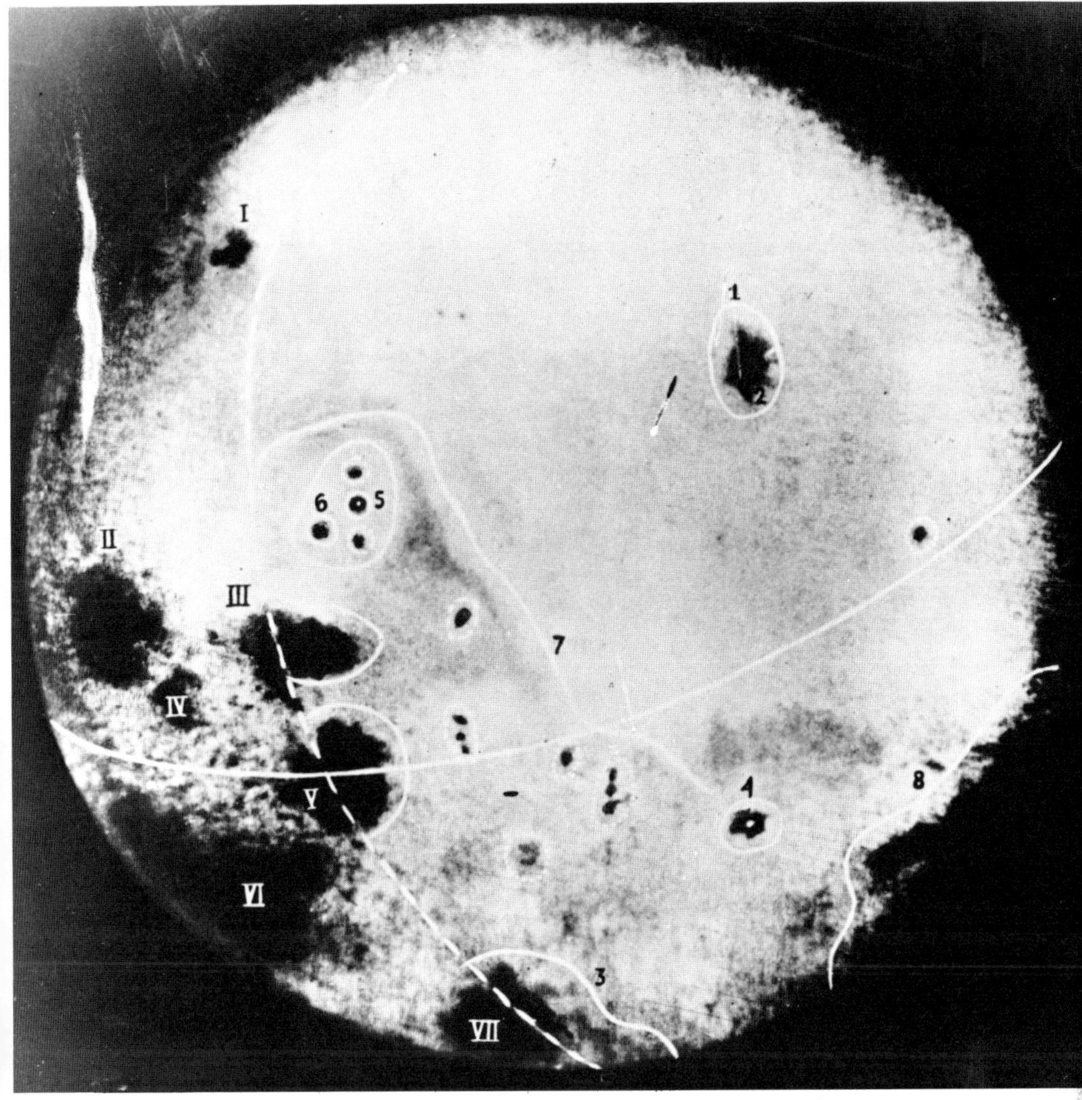

1 The Sea of Moscow, a 180-mile diameter crater area
2 Cosmonaut Bay in the Sea of Moscow
3 Continuation of the Southern Sea on the reverse side of the moon
4 Crater with Tsiolkovsky Hill
5 Crater with central Lomonosov Hill
6 Joliot-Curie Crater
7 Sovietsky mountain range
8 Sea of Dreams

The solid line is the lunar equator, and the dotted line is the boundary between the visible and the invisible sides of the moon as seen from earth.

The Roman numerals indicate objects on the visible part of the moon:

I	Humboldt Sea	V	Smith Sea
II	Sea of Crisis	VI	Sea of Fertility
III	Regional Sea	VII	South Sea
IV	Sea of Waves		

Left *A replica of Lunik 3 which first photographed the previously unseen side of the moon. The chequered panels are solar cells which powered the aerials (at the top) to transmit pictures from the camera (in the base) back to earth.*

Above *The back of the moon as photographed by Lunik 3.*

Left *The first man in space, Hero of the Soviet Union, Yuri Gagarin.*

Above *The recoverable capsule of Gagarin's spacecraft showing the charring caused by re-entry.*

The Space Race Kennedy commits the US to a moon landing in this decade

Left *The Vostok launcher, the key to Russia's successes in space. First revealed in 1967 at the Paris Air Show, when Western experts were astonished by the sophistication of the rocket and its vernier controls, clearly shown in this photograph.*

At the beginning of October 1958 the International Geophysical Year was drawing to a close: the Russians had now astonished the world with three highly successful Sputniks, and although the United States had replied by putting three objects into orbit, it was clear that the Russians were a long way ahead. The American space effort was still a matter of inter-service rivalry.

The US Navy was battling on with its ill-fated Vanguard programme. The Army, now permitted to use its Jupiter rocket for space exploration, had every intention of continuing its Explorer series. The Air Force, backing the Thor-Agena combination, was now ready to use it to place the first of its Discoverer satellites into Polar orbit.

It was against this background that the National Aeronautics and Space Administration was set up, with $330,000,000 for its first year's work. It was created as a deliberately non-military organisation; however nearly all its staff, premises, and certainly all the launch vehicles at its disposal, were of military origin. Its task was to initiate a logical space programme: to do so it would not only have to survey existing US projects, but also kill the inter-service rivalry which some thought had been instrumental in allowing Russia to take such a commanding lead in space technology.

Within a week a committee headed by Robert Gilruth presented a programme for putting the first American in space: Project Mercury. In March of that year a report had been published, entitled *Preliminary Studies of Manned Satellites, Wingless Configurations, Nonlifting*, under the aegis of the National Advisory Committee on Aeronautics (a precursor to NASA) high altitude rocket plane programme. Some of the initial work had already been done at the Langley Aeronautical Laboratory in Hampton, Virginia, and it was here that Gilruth came as head of the Space Task Group to finish the preliminary studies. America was entering the space age warily. NASA insisted that existing hardware should be used wherever possible, and that safety should be the overriding consideration: unlike the USSR they had opted to carry out their space programme without the protective shroud of secrecy that a programme with military elements might have made essential. Because of this any accident early in the project could have done irreparable damage to its public image.

This open policy, however public-spirited it may now appear, could have caused Congress to stop NASA's manned space projects if a disaster similar to the Apollo fire of 1967 (when Grissom, White and Chaffee were incinerated on the launch pad) had happened early in the Mercury or even the Gemini programmes. As it was NASA's publicity-conscious attitude may have been a blessing in disguise – not merely in propaganda terms, but in its approach to the role of the astronaut. Critics of the Russian programme say that their cosmonauts are merely standby pilots who take over if the automatic systems fail. The American astronaut is very much more an integral part of the control system of his spacecraft.

Even on Mercury, although NASA had chosen to make the flight automatic, they designed the guidance systems so that astronauts could have manual control at any time – a decision for which they were soon to be grateful.

By February 1959 NASA were ready to hand out the first contracts for building their space vehicles. The contract for the Mercury capsule went to McDonnell Aircraft, and the space race suddenly became an essential part of the American economy – although few, even at that time could have foreseen that, in less than ten years, 20,000 companies and 350,000 US citizens would be employed in the effort to put an American on the moon. NASA had not yet made a final decision on which rocket to use as the launch vehicle for its capsule, while the Russians had gained another space first with Lunik 1 (2 January 1959), the first man-made object to escape earth gravity.

The Americans had five rockets from which to choose. The only one specifically developed for space was the Navy's Vanguard: it had a bad record of failures, only slightly redeemed by two successful launches that year (Vanguard 2, 17 February; Vanguard 3, 18 September). In addition it was only capable of launching 56 lb. into orbit.

The Army's Jupiter C was available: it had launched five Explorer satellites that year (three successful, two abortive). Theoretically it was powerful enough for suborbital manned flights. Its more powerful offspring, Juno 2, was not yet ready. The Air Force had two rockets on offer: the Thor, and the Atlas, the latter being the first American ICBM, designed to deliver an atom bomb over a range of 9,000 miles, and capable of putting 9,000 lb. into earth orbit. The other Air Force ICBM, the Titan 1, had only received its first test in February of that year, and to select it could have meant delay. The Titan 2, which was later so successful in launching the Gemini series, had little in common with Titan 1, and was not ready until 1962.

The Atlas (Atlas D) was the choice, although for suborbital tests NASA used the forerunner of Jupiter, the Redstone IRBM – a direct descendant of the German V-2.

The other two essentials for manned space flight, apart from a launch vehicle and a spacecraft, were the setting up of a world-wide tracking network to both follow the astronaut and pick him up from anywhere on his orbital path, and the selection of the astronauts themselves. The stimulus of science fiction has played a remarkable role in furthering man's efforts to leave earth, but its effect on the young men of 1959 and 1960 could have been disastrous if they had read an advertisement saying 'Astronauts Wanted' – and the task of the selection committee would have become impossible. NASA narrowed the field by assembling an astronaut identikit. 'He' – and it is interesting that NASA, unlike the Russians, do not seem to have seriously considered the possibility of female astronauts – would be an 'all American' boy, in the peak of condition, in his mid-thirties (the ages of the 7 finally selected ranged between 32 and 38), exceptionally intelligent, cool, and determined. He would have to be capable of learning a new science, and of virtually writing his own instruction manual. To understand trajectories and Keplerian orbits he would ideally be an engineer with flying experience. The selection committee decided to limit the choice to test pilots from the armed services; a move to recruit pure scientists (following the Russian practice, and only adopted by the USA in 1965) was quashed. They strongly favoured married men, reflecting the American attitude that marriage is evidence of character stability; to this day only 3 of the 66 men qualified are unmarried. One psychological defect was particularly guarded against – the impulse to self-destruction, the 'death wish'.

In February 1959, by a combination of accepting volunteers and issuing orders, 508 supermen were assembled in Washington. The late Virgil 'Gus' Grissom recalled being put in a room with a score of other men, with strict orders not to reveal his job: 'I was convinced that somehow or other I had wandered right into the middle of a James Bond novel.' Medical tests and exhaustive questioning reduced the number to seven.

The list read:

Carpenter, M. S. (BSC)
Glenn, J. H. (DFC, AM)
Grissom, V. I.
Cooper, L. G. (BSC)
Schirra, W. M., Jr. (BSC)
Shepard, A. B., Jr. (BSC)
Slayton, D. K. (BSC)

All were married, and most of them had two children. Every one of them had the intelligence, enthusiasm and experience to be able to get into just about any flying-machine known to man, and master it within days. Although some were commanded to appear before the Committee, no one was finally ordered to take part in the Mercury programme – they were seven genuine volunteers.

First they were given what was probably the most thorough medical examination that has probably ever been suffered by any man. Potential danger points investigated included blood pressure, capillary resistance, eyesight (both normal and after dazzle), nocturnal vision, colour vision, acuteness of hearing, neurological symptoms, and respiration. Then the seven were sent to the Wright Patterson Military base in Ohio for further exhaustive tests, both in flight and in complex simulators, checking bodily reactions through every extreme it was thought they might one day have to experience. They suffered the rigours of low pressure chambers, high temperature chambers, and even isolation chambers to see if they developed claustrophobia, or hallucinations; they sat for hours with their feet in bowls of iced water, while teams of physicians recorded the effects on heart rate, breathing, metabolism, body temperature and oxygen consumption. They were given lectures by experts on anatomy, physiology and astronomy – and judging by the success of Apollo 8, and the respect which later astronauts like

Borman, Lovell, and Anders gained from astronomers, their lessons stood them in good stead for the real thing.

At the Pensacola Naval School of Aeronautical Medicine they were given refresher courses in night vision, and spent hours in another kind of isolation chamber – this time, a revolving one. Nausea is produced at between 5 and 10 revolutions a minute, and here again their sensitivity had to be assessed.

Perhaps the biggest problem facing the medical experts was weightlessness. At that date no one knew how a living body would react to a prolonged period of weightlessness: it is almost impossible to simulate it on earth. A whole series of harnesses and swivelling platforms was rigged up to give the astronauts a few seconds of experience at zero gravity. They spent hours immersed in tanks of water and were encouraged to take up skin-diving off the Florida coast to become familiar with three-dimensional free movement along three axes. For training they were taken up in aircraft which gave them a maximum 50 seconds of weightlessness by flying a parabolic course: roughly speaking the aircraft followed a similar path to a stone thrown in the air – decelerating on the way up, and accelerating on the way down. Such manoeuvres made it possible to practise only the briefest operations – like emergency exits from capsule doors or drinking a glass of water. Unfortunately the high acceleration needed both before and after such parabolic flights made them impractical for testing physiological changes, and the scientists had to wait for orbital conditions before they knew precisely what effect weightlessness would have. Many were apprehensive and expected minor disorders like anorexia (lack of appetite), nausea, sleeplessness, euphoria, gastro-intestinal disorder and even muscular atrophy. There was doubt too whether even the simple process of eating would be straightforward in space, without the constant pull of gravity to assist the swallowing and digestive processes. The physiologists could only guess at the effect of long periods of disorientation, when no stimuli originate in the canals of the inner ear to tell the body whether it is pointing up, down, or sideways.

There was one more related problem – high gravity, the 'g' forces with which every pilot is already familiar. Uncomfortable as they are, they posed relatively few problems as they are easier to simulate in centrifuges, the horror of all astronauts. Anyone who travels in a fast elevator experiences slight 'g' as it begins to rise; but the astronaut's description of 'eyeballs in' and 'eyeballs out' is some indication of the forces a would-be spaceman must undergo. The astronaut encounters 'g' both on launch and on re-entry. 'G' simply means the apparent multiplication of one's weight during the process of acceleration: thus the 6·8 g which the Mercury astronauts experienced two minutes after lift-off, meant that they felt they weighed 6·8 times more than normal. The highest 'g' forces were, however, experienced at the end of the flight during re-entry – 7·8 'g'. If astronauts were to attempt to withstand such forces in an upright position they would black-out at around 5 g: but lying on their backs reduced the amount of blood being forced out of the brain, and thus produced higher 'g' tolerance.

Training can greatly improve the ability to withstand 'plus g' – and because of this every astronaut spends many long, highly unpleasant hours being whirled around in the centrifuge. Only then can he guarantee that he will be able to perform simple actions in high gravity conditions – uncomplicated but vital movements like pressing the abort button if something goes wrong in the seconds after lift-off. During the centrifuge course the astronauts were often deliberately subjected to 'g' forces producing black-out. It is not for nothing that the Russian cosmonauts call the centrifuge the 'Devil's Mill'. Jim Lovell of the Apollo 8 crew told me that it was the worst experience of all his rigorous training programme.

From the very first stages of development of their spacecraft, the astronauts were intimately involved in its design and development, and even suggested changes and improvements. With their high level of technical knowledge, this was a feasible and no doubt beneficial exercise. They were given a course of back-up lectures on modern engineering techniques so that they could talk with the engineers designing the Mercury capsule at McDonnell's St Louis plant in their own language.

In parallel with the development of the Mercury capsule, a simulator was developed which would reproduce as faithfully as possible all the conditions of space flight. Computer controlled operational circuits monitored each astronaut's performance in manual control, and 'faults' could be introduced without warning, requiring prompt action from the astronaut. Each of the seven spent 90 hours in the simulator, and voice communications between them and the simulator controller were recorded, to be replayed again and again to Mercury Flight Control: complete familiarity with each astronaut's voice was essential in order to interpret every nuance as an indication of his precise degree of confidence.

Space navigation was taught in a planetarium, and they learnt everything from photography to jungle survival, to be prepared against every eventuality. Considering the complexity of the control panels and systems procedures inside not only Mercury, but also Gemini and Apollo capsules, many have puzzled how the astronauts remember everything: in this at least they are more fortunate, for on both real missions and mission simulation they carry a check list to guard against omitting some vital sequence.

While the engineering team in St Louis was producing, as far as they knew, the world's first spacecraft, the Space Task Group turned out a series of semi-functional replicas of the capsule for preliminary testing. Such replicas are known in NASA jargon as 'boilerplate' models: they are made of steel and often have the same dimensions and weight as the real thing, but lack the sophisticated equipment. They are used on all preliminary testing of launch vehicles, and the last one to fly was in December 1968 aboard Apollo 8 – a boilerplate equal to the weight of the lunar module carried on Apollo 9. The first boilerplate of the Mercury capsule was flown successfully in October 1959, launched by a cluster of Little Joe 1 rockets (the solid fuel rocket which has now been developed into the launch escape rocket for Apollo, sitting on top of the command module high above the Saturn V launch vehicle).

The first fully operational Mercury capsule off the McDonnell production line was delivered to the testing range at Wallops Island in Virginia on 12 April 1960. Considering it had been delivered only 14 months after the contract had been signed, it represented a remarkable feat of engineering. It was 11 feet tall and weighed about 4,000 lb. Its familiar, blunt-nosed, conical shape was the result of careful design so that it would suffer minimum punishment on re-entry. Its convex base was coated with a heat shield of glass fibre embedded in resin. As the temperature rose during re-entry, due to the friction of air moving past the capsule at 25,000 feet per second, the resin boiled off, keeping the skin of the shield at a maximum 3,000 deg. F., dissipating the heat and preventing it from passing to the astronaut chamber inside. During launch another heat problem was encountered by the alloy shingles on the cone, rising to 1,300 deg. F.: to radiate this back into the atmosphere the exterior was painted black.

Instead of the flaps and rudders of a conventional aircraft, the Mercury astronaut was equipped with 12 hydrogen peroxide thrusters with which he could control the pitch, roll and yaw of the spacecraft. He controlled these with his right hand using a kind of joystick, and his left hand covered the emergency escape control. Below the data recording equipment on his right lay the vital 'environmental control' apparatus, supplying both the cabin and his space suit with pure oxygen at a very low pressure, and at a temperature adjustable between 50 and 80 deg. F.

The choice of pure oxygen has raised many criticisms of the NASA programme, but there were good reasons: the enormous disadvantage is that it drastically increases the fire risk inside the capsule – with the tragic result in the Apollo programme of the death of Grissom, White and Chaffee on 27 January 1967. The disadvantage of using air is that the astronaut would get what deep sea divers call 'the bends' if the cabin depressurised for any reason. A compromise, often used under the sea, is an oxygen/helium mixture – but that causes communication problems: human speech, in those conditions, sounds remarkably like Donald Duck. The overwhelming advantage of pure oxygen was that pressure could safely be kept low at 5 psi (though only for short flights – and even then, the Mercury astronauts had to spend some time before each flight breathing oxygen enriched air at a higher pressure). Because oxygen meant low pressure inside the capsule there was no need to build a heavy pressure shell: for the Russians this was not so important as their rockets were more powerful and could easily cope with the extra weight of a strong pressurised capsule. For the Americans the weight factor was crucial, and oxygen was chosen. Stale air was extracted by a fan, body odours removed by activated charcoal, and carbon dioxide by lithium hydroxide; water was condensed into a sponge, which by periodic squeezing gave drinking water.

Three rockets accelerated the capsule away from its Atlas launcher after lift-off, and three more slowed it down before re-entry. The first mission 'flown' by a Mercury capsule was a test flight referred to as a 'beach abort', on 9 May 1960 at Wallops Island. This was a modest affair – an extension of the Little Joe cluster launches with boilerplate models – and was a complete success. Then the trouble started.

The Redstone rocket, selected for the ballistic testing of the Mercury capsule, grew temperamental and caused the failure of several projected missions. The plan for 8 sub-orbital flights was rapidly reviewed and cut down. On 29 July mission MA-1, the maiden flight of the Mercury-Atlas combination, was more or less a failure. In November 1960 Mission L-J 5 was equally disappointing. On the other side of the world Russia had launched the first two successful Sputniks in the 'Korabl' series, containing animals: and there were growing rumours that the USSR was about to attempt the first manned suborbital flight.

Then on 19 December 1960 the Mercury capsule was launched with modest but much needed success. It rose to 130 miles, and in the following January another Redstone lifted a capsule containing Ham, the first chimpanzee of space, for a 16-minute 39-second ride. The world's Press was delighted: Ham achieved an altitude of 157 miles, and his grimace of abject terror was misinterpreted by almost every journalist as a grin of happy delight. The Americans became more hopeful, and M-A 2, the first successful mission using the Atlas rocket, made a perfect ballistic flight with an unmanned, but heavily instrumented Mercury capsule on 21 February 1961.

On 12 April 1961 the engineers at Cape Canaveral came to work, preparing for the first orbital mission of the series, M-A 3. It was to be a normal working day, with the prospect of a successful unmanned Mercury orbit in the not too distant future: however on the radio they heard a Russian shouting 'I am an eagle', and the morale of the most technologically advanced nation on earth collapsed. Yuri Gagarin was completing his 89-minute orbit of the earth.

NASA still had three vital missions to accomplish before they could hope to put a man in orbit. The unmanned Mercury capsule still had to be orbited, then an astronaut had to be placed on a suborbital trajectory, and, finally perhaps ironically, 'Ere the leviathan could swim a league', a chimpanzee had to 'put a girdle round about the earth in forty minutes' – before Mercury was ready for manned orbit. NASA was committed to a policy of caution. It must have been a considerable temptation to skip the third mission, the chimpanzee orbit – the fact of Vostok 1's flight had proved that there was no absolute or unimagined bar to man's survival in space. However NASA decided to stick to its plan. But if the space race had gone unnoticed before, now it was open, and in earnest, and the USA was several laps behind.

Thirteen days after Gagarin's triumph, the Mercury capsule, mounted on the gigantic Atlas launch rocket, was fitted with its first passenger for an orbital mission. He was christened the 'Plastinaut' by the ground crews, an artificial astronaut designed to behave in certain respects like a human, to test both the life support system under full operational conditions, and to give the recovery fleet a full dress rehearsal. The countdown went without hitch, and ignition had just been achieved when Mission M-A 3 developed into one of the most spectacular failures of the entire series. Without even clearing its service tower the Atlas rocket hesitated, sank down a fraction, and disintegrated into a mammoth fireball as thousands of gallons of kerosene and liquid oxygen released in seconds the energy intended to put the Plastinaut into earth orbit.

The guidance system had failed, and the Range Safety Officer had pressed the button marked 'destruct'.

The loss of the $7,000,000 Atlas rocket may have seemed disastrous at the time, but NASA was lucky not to lose an astronaut as well in that particular holocaust: flying 1,000 feet above the launch pad, observing the launch, was astronaut Virgil Grissom. From the ground it looked as though he had flown straight into the fireball. He coolly watched the ejected Mercury capsule float down to the sea, and only later realised that the Plastinaut in it (the launch escape tower having functioned to perfection) was much safer than he himself. Grissom recalls thinking, as he watched, what big seagulls were flying round the Cape that day. Only when some of the 'seagulls' overtook his delta wing F-106A at high speed did he realise that they were chunks of shrapnel from the Atlas rocket. Miraculously he was not hurt.

It was decided to proceed with M-R 3, a manned ballistic flight. The date was fixed for 2 May, at 3.30 a.m. Cape Canaveral was immediately besieged by the Press, all wanting to know the name of America's first spaceman. John Glenn was the favourite, but not until late on 1 May did they announce that it was to be 38-year-old, ex-Naval test pilot, Commander Alan B. Shepard, Jnr. Glenn had done better than Shepard in tests, but NASA had decided to reserve him for the later and more arduous, first orbital flight.

Weather caused a 3-day postponement, but on 5 May 1961 a jubilant Shepard was lifted to an altitude of 114 miles, reaching a maximum speed of 5,188 mph. He can only have been above the atmosphere for a few minutes, and as an achievement it could not compare with the Gagarin orbit. But a little prestige had been regained – an American had entered space.

The pioneers of the space programme in the United States had been lobbying for some time for a real commitment to manned flight which would lead one day to putting a man on the moon. However, Congress was cautious, balancing the huge sums of money required against the early failures in the

Mercury programme. The public generally saw it all as a race, although some critics were already pressing for unmanned, robot space exploration as opposed to the expensive and much more dangerous manned missions. The space programme cost the US over $900,000,000 for the 1961 financial year. If Shepard's flight had been unsuccessful Congress might well have forced NASA to lower its sights and accept more modest objectives.

Twenty days after Shepard's brief excursion into space President John F. Kennedy made one of the most significant speeches of his whole career. It was to commit the United States to getting an American on the moon within the decade. The occasion was the opening of a debate in Congress on Defence and Armaments. He said:

> If we are to win the battle that is now going on around the world between freedom and tyranny, the dramatic achievements in space which occurred in recent weeks should have made clear to us all, as did the Sputnik in 1957, the impact of this adventure on the minds of men everywhere, who are attempting to make a determination of which road they should take. Since early in my term, our efforts in space have been under review. With the advice of the Vice-President, who is Chairman of the National Space Council, we have examined where we are strong and where we are not, where we may succeed and where we may not. Now it is time to take longer strides – time for a great new American enterprise – time for this nation to take a clearly leading role in space achievement, which in many ways may hold the key to our future on earth.
>
> I believe we possess all the resources and talents necessary. But the facts of the matter are that we have never made the national decisions or marshalled the national resources required for such leadership. We have never specified long-range goals on an urgent time schedule, or managed our resources and our time so as to insure their fulfilment.
>
> Recognising the head start obtained by the Soviets with their large rocket engines, which gives them many months of lead-time, and recognising the likelihood that they will exploit this lead for some time to come in still more impressive successes, we nevertheless are required to make new efforts on our own. For while we cannot guarantee that we shall one day be first, we can guarantee that any failure to make this effort will make us last. We take an additional risk by making it in full view of the world, but as shown by the feat of astronaut Shepard, this very risk enhances our stature when we are successful. But this is not merely a race. Space is open to us now; and our eagerness to share its meaning is not governed by the efforts of others. We go into space because whatever mankind must undertake, free men must fully share.

This was the President of the United States speaking openly of the space effort in terms of a race with the USSR, a propaganda race, a prestige struggle between, as he saw it, 'freedom and tyranny'.

> First, I believe that this nation should commit itself to achieving the goal, before this decade is out, of landing a man on the moon and returning him safely to the earth. No single space project in this period will be more impressive to mankind, or more important for the long-range exploration of space; and none will be so difficult or expensive to accomplish. We propose to accelerate the development of the appropriate lunar space craft. We propose to develop alternate liquid and solid fuel boosters, much larger than any now being developed, until certain which is superior. We propose additional funds for other engine development and for unmanned explorations – explorations which are particularly important for one purpose which this nation will never overlook; the survival of the man who first makes this daring flight. But in a very real sense, it will not be one man going to the moon – if we make this judgement affirmatively, it will be an entire nation. For all of us must work to put him there.
>
> Secondly, an additional 23 million dollars, together with 7 million dollars already available, will accelerate development of the Rover nuclear rocket. This gives promise of some day providing a means for even more exciting and ambitious exploration of space, perhaps beyond the moon, perhaps to the very end of the solar system itself.
>
> Third, an additional 50 million dollars will make the most of our present leadership, by accelerating the use of space satellites for world-wide communications.
>
> Fourth, an additional 75 million dollars – of which 53 million dollars is for the Weather Bureau – will help give us at the earliest possible time a satellite system for world-wide weather observation.
>
> Let it be clear – and this is a judgment which the Members of the Congress must finally make – let it be clear that I am asking the Congress and the country to accept a firm commitment to a new course of action – a course which will last for many years and carry very heavy costs; 531 million dollars in fiscal '62 – an estimated seven to nine billion dollars additional over the next five years. If we are to go only half way, or reduce our sights in the face of difficulty, in my judgment it would be better not to go at all.
>
> It is a most important decision that we make as a nation. But all of you have lived through the last four years and have seen the significance of space and the adventures in space, and no one can predict with certainty what the ultimate meaning will be of mastery of space.
>
> I believe we should go to the moon.

The timing of the speech was masterly. The nation was still delighted over Shepard's success, but had not forgotten Gagarin. Public opinion was behind Kennedy, Congress gave in, and the race became official.

An *ad hoc* Lunar Study Committee was set up to study details. The Apollo project for putting man on the moon had been tentatively announced back in 1960 to an Industrial Press Conference. A manned expedition to the moon before 1970 had been recommended by the Government and Space committee on 2 July 1960. By 15 July 1961, less than two months after Kennedy's speech, NASA was ready to invite tenders for the spacecraft that was to take three Americans to the moon before 1970.

The United States had embarked on the most ambitious project in its history, and the success of the Mercury programme was more than ever vital. On 21 July, 'Gus' Grissom flew a successful but again suborbital flight on mission M-R 4. He called his craft Liberty Bell, and it gave the rescue forces a nasty moment by sinking on splashdown. The escape hatch blew open prematurely, and Grissom was only saved from drowning by a 'neck dam' that Schirra had persuaded him to wear: a watertight ring inside his spacesuit that prevented it from filling with water.

Liberty Bell had sunk before retrieval after a simple ballistic flight, and there was now growing concern over the fate of orbital astronauts who could be coming down in much less predictable places. Then on 6 August this concern turned once more into national dismay: Titov orbited the earth 17 times in Vostok 2. It was not until 29 November that a chimpanzee, Enos, orbited the earth three times after being launched by M-A 5, and cleared the way for Mercury's primary objective, a manned orbit.

Meanwhile the medical experts had heard of Titov's 'space sickness', and were beginning to fear that weightlessness might, after all, be the major hazard of space travel. Finally, on 20 February 1962 Mercury Atlas 6 soared up into the clear Florida sky, and within minutes the world knew that John Glenn was in orbit. The orbit was designed so that the spacecraft would naturally spiral in and re-enter after 7 orbits, even if all control over it was lost. He called his craft Friendship 7, and she travelled at 25,730 feet per second, reaching an orbital peak (apogee) of 162 miles from earth, and a perigee (the lowest orbital altitude) of 100 miles.

The flight lasted 4 hours 55 minutes; at the end of only three orbits it was decided to return to the ground, and it was at this point that the flight controllers on the ground were thankful that the astronaut himself could independently override the automatic control systems. Pre-entry checks revealed that the auto-control system governing the retro-rockets (which would brake the capsule out of its orbital path) were not functioning correctly. Taking over control was an exercise which Glenn had performed scores of times back on earth in the safety of the simulator. Calmly he ignited the retro-rockets himself to begin the spacecraft's curving descent back through the atmosphere, splashing down 166 miles east of Grand Turk Island, in the Bahamas.

The Mercury programme had achieved its objective and

Colonel Glenn received the hero's welcome he deserved. Apprehensively NASA awaited the Russian reply, fearing another prestigious feat by way of propaganda retaliation. All they received was the usual courteous cable of congratulations on 'a fine achievement'. Mercury continued with Malcolm Scott-Carpenter's flight in Aurora 7 (M-A 7) on 24 May. Again there were agonising launch delays, before what was to be a good flight involving slightly more complex experiments, and a cliff-hanging end when the capsule landed 250 miles off target and was not located for some hours.

Even before Colonel Glenn had gone into orbit another NASA project had been announced – Mercury Mark 2 – which was to develop into the Gemini programme using two astronauts. By now every decision was becoming inextricably involved with the early planning of the lunar landing expedition, Apollo itself. At that time it seemed there were four practical methods of getting to the moon.

The first was the direct one: a Command Module, Service Module and Lunar Excursion Module could be launched together towards the moon as a 3-module complex. The LEM would make in-course flight corrections and decelerate all three for a combined lunar landing. On return the LEM would be left behind and the SM would become the launch rocket for the return journey. Finally the CM containing the astronauts would jettison the SM just before re-entry to earth atmosphere, and become the only part to return to earth.

The second was the Earth Orbiting Rendezvous method (EOR). An unmanned satellite, basically a 'tanker' vessel full of space fuel, would be placed into earth orbit. A similar 3-module complex, carrying the minimum of fuel, and therefore much lighter, would be launched and docked with the orbiting tanker for the transfer of fuel. The 3-module complex would then accelerate to the moon, performing the same manoeuvres as in the direct approach.

The Lunar Orbiting Rendezvous technique (LOR), the third method, meant putting the CM, LEM and SM into lunar orbit. Two astronauts would descend to the moon in the LEM leaving the CM and SM in lunar orbit. Then half the LEM would lift off the moon surface, using the other half as a launch pad. The crew would rejoin the astronaut who had remained in the orbiting CM and use the SM to propel them back to earth, jettisoning it before re-entry.

Finally they considered the Lunar Surface Rendezvous: here a service module, fully fuelled, would be soft-landed on the moon, by remote control. The astronauts would travel to the moon in a command module and lunar excursion module, and use the service module as their launch vehicle for return to earth.

The disadvantage of the direct approach is that to launch three modules, fully fuelled, and accelerate them to escape velocity would require an immense rocket. The most powerful rocket that NASA expected to have for the Apollo trip, the Saturn V, would have been far too weak. Nevertheless NASA commissioned a design for such a rocket: it is called the Nova, with a thrust of 5,600 tons at lift-off, capable of orbiting 200 tons (cf. Saturn V 130 tons). It has been on the drawing boards ever since, and has now been shelved indefinitely.

Because of expense and technical difficulties in orbiting large amounts of fuel for long periods the EOR method was dropped. LOR seemed the cheapest, but not the safest. The complex rendezvous requirements which would have to take place 240,000 miles away from Mission Control at Houston left frightening margins for error – both human and mechanical. It is easy for us with the conceit of hindsight to look back at the successful Apollo mission around the moon of Christmas 1968 and wonder why, less than a decade before, men had feared the wisdom of such techniques: yet, at that time, the fact that the two vital steps of LOR (firing the Service Propulsion System to brake into lunar orbit, and re-igniting to accelerate for the return journey) would have to be performed out of touch with earth, on the unseen side of the moon, made the method seem too hazardous.

NASA was first inclined to adopt the Direct Approach; it was then realised that Nova could not possibly be ready until at least two years after Saturn V. LSR was seriously considered, but dismissed for the initial trip to the moon because of the possible human difficulties, once on the lunar surface, of preparing the awaiting service module for launch. In July 1962 the decision was finally taken that three astronauts would go to the moon by LOR.

The Russians are believed to have considered both the Direct Approach and EOR, and to have chosen the latter. They revealed their interest in bringing spacecraft together by flying Vostok 3 and 4 within about 4 miles of one another. Cosmonauts Andrian Nikolayev and Pavel Popovitch were launched on 11 and 12 August, and the feat of accuracy of their being placed into a virtually identical orbit not only impressed the world yet again, but proved that docking in space was not far away. American accuracy was also improving: on 3 October Walter Schirra performed a textbook flight, taking colour photographs of the earth, making astronomical observations, and splashing down after 6 orbits and a flight of 9 hours 13 minutes only 5 miles from the recovery ship.

Nikolayev in Vostok 3 had the duration record of that time for space flight of 84 hours and 22 minutes, but the final Mercury mission (M-A 9) of 'Gordo' Cooper on 15 May 1963 was also a marathon – 22 orbits in 34 hours 19 minutes. The flights of Vostok 3 and 4, and M-A 9 proved to the Russians and Americans that there was no immediately obvious barrier to man's survival in space for extended periods, and the Mercury programme had finally proved a total success.

A week after the final flight, a letter arrived at Cape Kennedy from a motor mechanic in Milwaukee. It read: 'I see from my newspaper today that the Mercury programme cost 2 dollars and 6 cents for every man, woman, and child in the United States of America. I enclose a mail order for $2.06, and I'd like to see the whole thing over again.'

Right *The launch of Mercury Atlas 9, carrying Gordon Cooper into 22·9 e[…] orbits on 15 May 1963.*

Man Walks in Space Leonov and White enjoy EVA

After Gagarin's historic earth orbit the Russians looked firmly established in their lead in the space race. Moreover they were pushing ahead. Gherman Titov's long mission of 17 orbits – 25 hours 18 minutes in space – was by no means simply an extended version of Gagarin's flight.

Nine hours after lifting off from the launch pad at Tyuratam in Vostok 2 Titov reported growing sensations of both nausea and disorientation. His instrument panel appeared to 'float above his head', and remained suspended in space. These were the typical symptoms of disorientation which both Russian and American doctors had already feared would result from prolonged periods of weightlessness. Soviet flight control were tempted to call the mission off. Titov, however, insisted that he felt able to fulfil his mission, so the flight ran its full term. Twice he succeeded in taking over manual control, and he became the first film cameraman in space. The prime objective of this flight was to investigate weightlessness, and it was an ironic coincidence that this should be the only flight to date (both Russian and American), where such severe ill effects have been experienced.

After Vostok 2 there was a gap of a year during which no Soviet manned flights were announced. It has been suggested that there may have been some failures, but this can only be pure speculation. It is very possible, however, that the Russians were highly concerned about Titov's nausea, and, for once, were letting the USA catch up a little to see if American astronauts experienced similar trouble on long flights. Scott Carpenter's uneventful three orbits in May '62 could have been the extra reassurance the Russians needed, for only two months later on 11 August Vostok 3 carried cosmonaut Andrian G. Nikolayev into the first of 64 orbits. The next day he was followed into space by Vostok 4 carrying a pilot, Pavel R. Popovitch.

Russia seemed to be investigating the possibility of a rendezvous between two craft in orbit: only when this was achieved could any construction work, like the erection of an orbiting space station, be carried out in space. Also, as we have seen, it is an essential preface to both the EOR and LOR techniques of lunar landing. Vostok 3 and 4 had no facilities for actual docking, but the achievement of proximity was quite remarkable. Vostok 4 initially established a very similar orbit to Vostok 3 – slightly larger, and therefore slower, and inclined by a fraction of a degree to that of Vostok 3. These differences meant that, as Popovitch regained full control of himself after the high 'g' of insertion into orbit, he saw some three and a half miles ahead and slightly to the left his sister ship Vostok 4. It oscillated from left to right across his field of vision. Because of the two-mile difference in the perigee and apogee it also appeared to rise and fall on a vertical axis. Finally, because Vostok 4 was on a slightly slower orbit, the distance between the two increased until they lost sight of one another. No attempt was made to bring them any closer together, but they achieved the first VHF radio contact between two spaceships, and practised attitude changes.

Rendezvous of orbiting spaceships is a very sophisticated operation. For one spaceship to catch up with another it ignites a rocket to thrust it away from its target: this slows it down and it begins to drop under the influence of earth gravity. The target recedes. However, once the rocket is turned off, the vehicle is once more in free orbit, and it increases speed due to being in a lower, faster orbit. If the rocket burn has been accurately calculated the two spacecraft meet half an orbit later at the other side of the world.

If the spacecraft had directed itself towards the target in the first place, it would have travelled in a straight line until the rocket motors were turned off. Then it would have been further from earth, in a higher and therefore slower orbit, and would have fallen behind the spacecraft ahead even more rapidly.

In practice, docking is achieved by a whole series of such operations, each delicately gauged to minimise wastage of fuel. The calculations needed to estimate the exact amount of rocket impulse needed demand a computer.

Popovitch and Nikolayev were not riding spacecraft equipped for such complex tasks, but the accuracy with which Vostok 4 was launched to within 4 miles of Vostok 3 represented a space bull's eye.

All the Vostoks were placed in what is called 'a naturally decaying orbit': if control had been lost they would gradually have been slowed down until they re-entered the atmosphere of their own accord: otherwise they would have remained in orbit until a rescue rocket could rendezvous and extricate them – assuming a rescue rocket existed and the life support systems lasted until its arrival (both somewhat unlikely eventualities at the time).

Neither Popovitch nor Nikolayev suffered any 'space sickness' although they unstrapped themselves from their anti-'g' couches and floated freely around their cabin.

Instead of eating food squeezed from plastic containers they took up a menu ranging from chicken, veal, and sandwiches to fruit cake and sweets – all in bite-size pieces – as well as water, coffee, and fruit juice. Vostok 3 and 4 were the first proof that reasonably normal food could be eaten in space. The Americans still reconstitute dehydrated food in plastic bags, using a water gun. The danger of normal food is that crumbs can fall loose and drift into inaccessible, and perhaps dangerous places during weightlessness. The risk is small, but the Americans feel it is not worth taking: that was why a later astronaut, John Young, received a severe reprimand for taking a corned-beef sandwich with him into space. Because much of an astronaut's space life is sedentary (apart from the essential exercise periods to keep reasonably fit) his daily calorific intake averages 2,500 calories. On earth a manual labourer gets through about 3,000 per day. Astronauts seem to enjoy their food, although the preparation and stowing of waste (the Americans release a germicide into the plastic bags to guard against bacterial growth) must become a chore at times. Jim

Lovell, before Gemini 12, expressed a slight liking for fruit cake – with the result that he had fruit cake every day of the flight: before Apollo 8 he swore he'd never suggest a preference again. On the Vostok series they were still pioneering, and worried about the potential build up of radioactivity in the food – a fear which fortunately proved to be groundless.

Vostok 3 and 4 carried various biological specimens and much of the cosmonaut's time was spent studying their reactions and carrying out experiments on them. They filmed the earth, stars, planets, and cloud formations, and Popovitch carried out experiments on the behaviour of liquids in weightlessness.

After the success of this flight there was silence from Russia. For ten months Western experts speculated on what their next triumph would entail. Surprisingly it was virtually a repeat performance of Vostok 3 and 4's flight, with only one headline-catching difference: Vostok 6 was piloted by the first woman cosmonaut, Valentina Tereshkova. Women may well play a major part in the possible colonisation of space at the end of this century, but there seems as little scientific reason today as in 1963 for selecting one to fill a cosmonaut role. One conceivable reason, considering she later married cosmonaut No. 3, Andrian Nikolayev, is the improbable and highly unromantic one that their children might prove interesting subjects for physiological study: the extremity of such an Orwellian proposition forces one to the only other logical conclusion: after ten months the USSR was in need of good publicity. For once they were in the position in which the Americans so often found themselves: they needed a space success for prestige reasons. Both Tereshkova and Bykovsky, the pilot of Vostok 5, withstood the flight very well: her part was planned for a duration of only 24 hours with an option to extend it to three days, and she was fit enough to take that option.

If publicity attraction was her real function, she fulfilled it with a vengeance, filling acres of the world's newsprint, with her post-flight world tour and her cosmonaut wedding.

The major scientific interest of the flight of Vostok 5 and 6 was the fact that Bykovsky, after five days in space, declared that he had become completely accustomed to weightlessness after the fourth day.

Both the US and the USSR had now completed their initial manned space flight programmes, Mercury and Vostok. The Russians were unquestionably in the lead. Neither country was yet ready or even capable of sending a man to the moon, and the Russians were even denying that it was in any way a Soviet objective. The strategy of the Soviet programme was, and is, obscure; however it was clear that both nations now needed to develop rockets of intermediate power and to learn how to use them, before moving on to the heavy artillery of the lunar brigade.

For the United States the intermediate stage meant the Gemini programme: its objectives were to orbit a two-man spacecraft, to solve the problems of linking two spacecraft (docking), and to discover whether man, clad in a space-suit, could survive extravehicular activity, or walking in space (EVA). EVA is not an essential part of the American moon landing programmes, but knowledge and experience of it was considered useful research for other future projects, and it could be essential in the building of a manned orbiting laboratory.

NASA now needed a reliable rocket capable of putting some 8,000 lb. into orbit in order to launch Gemini. As less heavy nuclear weapons were being developed with correspondingly smaller rockets there were no longer any plans for mammoth military launch vehicles: NASA had to carry out its own development work.

For the first time a rocket using liquid hydrogen as a fuel (in theory the most efficient) had been developed. It was called the Centaur and was due to be flown in combination with the Atlas missile in May 1962. It would have had sufficient power to launch Gemini; liquid hydrogen fuels were already planned for the monster Saturn rocket, and it was tempting to choose the Atlas-Centaur as the launch vehicle for the Gemini programme.

However, in late 1962, the safest course available was taken, and the Air Force's ICBM, the Titan 2, was selected. The Atlas-Centaur filled an equally important role in man's conquest of the moon by being used for Surveyor (see Chapter 10). Titan 2 completed its tests in March of 1962, and had a safety advantage over most other launch vehicles. To serve as an ICBM it had to stand primed with fuel, instantly ready to be fired in case of attack. Rockets like the Atlas took some time to be filled, and, because liquid oxygen boils fiercely at normal temperatures, were not only dangerously explosive, but prone to heavy icing on the outside, and evaporation problems. Titan 1 also used kerosene and liquid oxygen as fuel, but Titan 2 used a fuel ('Aerozine 50') and an oxidiser (nitrogen tetroxide) which exist at normal temperatures, are not explosive, and yet ignite spontaneously on contact.

To adapt the Titan 2 to the Gemini programme several modifications were made including the introduction of a malfunction detection system: this continually monitors the state of every vital component and circuit, sounding the alarm in both the spacecraft and at Flight Control Centre if anything goes wrong.

Outwardly the Gemini spacecraft looked similar to, but slightly larger than, Mercury: as one of the astronauts put it, 'like a Mercury capsule that threw the diet rules away'. Because Titan 2 fuel was not explosive there was no need for the launch escape system used on both Mercury and Apollo to lift the capsule away from any emergency on the launch pad. Ejection seats were sufficient to guarantee astronaut safety – although during one ground test of the seats the hatch failed to open and the seat rocketed out through a closed hatch. John Young, who was watching, commented: 'One hell of a headache . . . but a short one'.

Both pilots could operate the controls, but unlike the Mercury capsule Gemini was provided with an on-board computer to do the complex rendezvous guidance calculations. Also, for the first time, it was possible to exert some control over the spacecraft once it had re-entered the atmosphere.

Mercury relied entirely on atmospheric drag to slow it down on re-entry, with no possibility of control. The Gemini capsule had a certain amount of aerodynamic lift, sufficient to carry it 350 miles further down range of the point where it would have landed on a normal ballistic path. Using 16 liquid propellant rockets, this lift could also be controlled to achieve a landing 300 miles short of the ballistic landing point, and 50 miles to right or left of it.

Two more groups of astronauts were selected for Gemini, and the training was even more intense, and technical. This time each astronaut specialised on one aspect of space flight, and then lectured his colleagues about it.

Only two unmanned test flights were thought necessary for Gemini, and the first, after many delays, was launched with complete success (G-T 1, 8 April '64). G-T 2, a Plastinaut flight, was scheduled for August, and the Titan rocket arrived at the Cape on time. Strike action then delayed its check-out, and when at last it was ready news came of Hurricane Cleo, heading straight for the Florida coast. Titan 2 was dismantled for fear of damage. Once re-erection was complete another forecast warned of Hurricane Dora. Again the rocket was dismantled.

The result of all these delays was that before Gemini could be put into orbit, the Russians, on 12 October, successfully launched their massive Voskhod 1. Inside the Russian craft were three spacemen, and only one of them, the late Vladimir Komarov, was a trained cosmonaut. Konstantin Feoktistov was a civilian scientist and Boris Yegorov an Air Force doctor.

The launch vehicle for Voskhod 1 was thought to be basically the same as that used for the Vostok series, but with an extra stage added. It boosted a spacecraft weighing more than 5 tons into an unusually high orbit (apogee 254 miles, perigee 111), and it was not a naturally decaying orbit as on previous missions. For safety Voskhod 1 had a standby retro-rocket as well as the main one, to initiate re-entry.

The Russians were so confident of the pressure shell and environment control system of their spacecraft that the three men inside wore no spacesuits. The cabin atmosphere was like that of a normal airliner.

Feoktistov spent most of his time making observations on the optical properties of the upper atmosphere; he also did some invaluable ground work on interplanetary navigation. He discovered for the first time exactly how the planet Earth could be used as a 'base line' for navigators travelling to Mars or Venus. For this reason some sources argue that Voskhod was a rehearsal for a Mars trip, rather than an equivalent of Gemini.

Dr Yegorov investigated space health, and tested his fellow crew members for their adaptability to weightlessness, their movement coordination, efficiency, respiration, and mental acuteness. American astronauts today are said to be enthusiastic about the Russian system of flying scientists as integral crew members. They tend to find the scientific experiments they have to carry out somewhat irksome, and would prefer to concentrate on their own specialist tasks. The crew of Apollo 7, who made headlines with their objections to acting simultaneously as TV stars and cameramen, also said they were overloaded with experiments extraneous to their actual mission. Perhaps to combat this feeling NASA now has 14 'Scientist astronauts', though none has yet flown.

Voskhod 1 stayed in orbit for 23 hours 17 minutes, circling the earth 16 times. The amount of scientific and military information collected was colossal. The craft was fitted with not only the usual internal TV cameras, but an additional one outside, and all of them relayed live pictures back to earth.

Another advance in Voskhod 1 was the use, for the first time in space, of an experimental type of electric rocket, an ion propulsion device (cf. final chapter), here used for attitude control.

Voskhod made an accurate landing on schedule, and chalked up another technological 'first' as it did. Because of the USSR's vast land mass, and the temperature of the oceans that surround it, the Russians have always preferred to bring their cosmonauts down on land, rather than using the American 'splashdown' method. In the Vostok series, after initiating re-entry, and releasing parachutes, the cosmonaut would fire his ejection seat at 22,000 feet and land independently. At the end of the flight of Voskhod 1 they radically improved the technique. The spacecraft descended to within 1,000 feet of the earth's surface, and then, with the crew still inside, a special soft-landing system was activated bringing the spacecraft to a 'feather-bed landing'. Quite how this was done remains a mystery, and the USSR gives no details beyond saying: 'Gunpowder devices were ignited': however it must be assumed that the craft was equipped with special retro-rockets for the purpose.

Back at Cape Kennedy NASA seemed to be not only in a race with the USSR, but with Fate herself. With the hurricanes and strikes over, Titan 2 was once more set up and checked out, ready for launch on 9 December 1964. The Press had almost lost interest, but the atmosphere at Cape Kennedy was tense, and Gemini 2 was already being called a jinx mission.

Countdown went smoothly, with only minor hold-ups. The 430,000 lb. thrust of the Titan's first stage motors roared into life. The cameras tilted towards the sky, then slowly panned back to reveal G-T 2 standing upright on its launch pad, not having moved an inch. The clamps, which hold the rocket down until its motors achieve maximum power, remained firmly in place.

The hydraulic system controlling the directional thrust of No. 2 engine had developed a fault: a standby system had been switched on, and the fault corrected. However the Titan's control system had a built-in instruction: in the event of a standby system being activated before launch, the rocket should not take off. In 15 milliseconds the new Malfunction Detection System had assessed the situation and shut off the engine.

Until that day many critics had argued that the MDS probably would not work, and even if it did, would be too expensive and too heavy. The MDS had now proved itself: damage to the rocket was only slight, and the efficiency of the new system gave everyone a much needed boost of confidence.

Finally, on 19 January 1965, Gemini 2 was launched into a suborbital flight carrying two Plastinauts 2,150 miles in 19 minutes, reaching a speed of 16,500 mph. After an intentionally harsher re-entry than was planned for manned flights, it was recovered two hours later by the aircraft carrier *Lake Champlain*, and the capsule was declared fit for human habitation.

The regularity with which Russia stole America's thunder at this time makes it difficult to believe that it was by pure coincidence. Gemini 3, the first manned Gemini mission, was scheduled for 23 March 1965. On the 15th of March the astronauts selected for this flight, 'Gus' Grissom and John Young, moved into the new astronaut quarters on Merritt Island. The countdown had already begun. Three days later there were two men in space: not Grissom and Young in Gemini 3, but Pavel Belyayev and Alexis Leonov in Voskhod 2.

Voskhod 2 followed an almost identical orbit to Voskhod 1, but carried only a crew of two, this time both fully trained cosmonauts, and both wearing spacesuits. Although it only carried two men Voskhod 2 was 205 lb. heavier than its forerunner: the extra weight was accounted for by a special, electrically-operated airlock which had been bolted to the hull of the spacecraft. It was through this airlock, during the second orbit, that the world saw Colonel Leonov climb: protected only by his spacesuit, carrying an autonomous life-support system, and attached to the craft by only a slender tether he pushed off to become the first man to float in space. He spent 12 minutes outside his capsule, and having attached a camera to the outside wall to record man's first encounter with the void, he floated away until he was about 15 feet from the spacecraft.

> With a light push I moved away from the spacecraft: I moved further and further from it until the line was all out. The small thrust I had given as I left it had imparted a slight angular motion to the capsule and I saw our wonderful vehicle rotating slowly before me. I had expected to find sharp contrasts of light and shade, but the reality was quite different: the sun's rays, reflected from the earth, illuminated the shadowed side of the capsule clearly.
>
> The vast cosmos was visible to me in all its indescribable beauty. My first glance was towards the earth, which seemed to move slowly past me. Despite the thick glasses of my helmet, I could see clouds to the right, the Black Sea below my feet, the Bay of Novorossysk, and beyond the coastline, the mountain chain of the Caucasus.

Pulling gently on the line, I began to draw myself back to the spacecraft, then, pushing off again and turning round, I moved away again slowly. A magnificent picture was spread out before me. I could see both the steady brilliance of the stars scattered over a background of black velvet, and at the same time the surface of the earth. I was flying over vast green masses: I recognised the Volga, the snowy line of the Urals, and the great Siberian rivers Obi and Yenisei, just as though I had been floating over a great coloured map. One could hardly imagine a more fascinating sight for an artist. In the black sky the sun shone brilliantly, and I felt its warmth on my face through my helmet window.

After a while, I pulled the line harder and saw the capsule rushing towards me. The first thing I thought of was the danger of striking my helmet window against the hull, but as I reached the airlock panel, I was easily able to fend it off with my hands. I realised that, with training, one could easily move about in these curious conditions with certainty and precision. I felt well, my nerves were steady and I had not the least desire to return on board. Even after I received orders to return, I moved away from the capsule once more to check the rotary motion which this action caused.

Finally I unscrewed the external camera which had filmed my space walk and got back through the airlock, with some difficulty because my space suit interfered with my movements.

I should like to draw certain conclusions from this experiment. It is perfectly possible to go into space and there is no longer any mystery about it. Wearing a special suit, man can not only survive, but also carry out coordinated movements and precise operations. It is just as possible to perform manual work in space as to make scientific observations.

This first space walk probably captured more public admiration than even Gagarin's achievement: the science fiction aspect of previous flights made them almost unbelievable, but now Moscow television had shown a man actually stepping into the unknown. Yet despite this extraordinary feat, the flight itself was only a partial success: it had been planned to last much longer, and Leonov's space walk was not its sole purpose. On the seventeenth orbit they had technical trouble. A fault developed in the auto-orientation system, making it impossible to bring the craft back to earth under remote control. The pilot Belyayev, having already showed remarkable skill in stabilising the craft during Leonov's walk, volunteered to bring her down manually. Ground control agreed, and re-entry began immediately. Belyayev controlled the descent impeccably, the soft-landing system worked, and Voskhod 2 ended its flight in deep snow, far to the west of the original landing area. Two hours later the cosmonauts were safely picked up by a recovery helicopter.

The overall Russian strategy was now becoming slightly less obscure. Wherever they finally intended to go, whether it was to the moon, to Mars or even to Venus, they intended to use a multi-manned mission. But Russia had still shown no interest in manned docking, and a pattern was emerging (especially in view of the 'Polyot' and 'Electron' satellite series which had come between Vostok and Voskhod) which showed a radical difference of approach.

Whereas Apollo astronauts were to be, in a very real sense, the indispensable pilots of their spaceships, it became clear that Russian cosmonauts, provided all went well, could be virtually passengers aboard fully automatic, remote controlled craft. At the same time Belyayev's manual take-over of the Voskhod 2 landing showed that the cosmonauts would be fully capable of taking control when necessary.

This general trend in Russian attitudes towards the control of spacecraft was confirmed last year when Academician Petrov, Director of the Soviet Academy of Science's Institute for Cosmic Experiments, criticised the American Apollo 8 'Christmas-round-the-moon' mission for excessive reliance on pilot skill as opposed to automatic operation.

With America firmly committed to a moon landing by 1970, controversy grew about Russia's intentions in space. Although an expedition to Mars might be considerably more worthwhile from a scientific viewpoint, its military value would be almost nil, and the Russian space programme has always been intimately tied to military planning.

In both countries there has always been opposition to a manned exploration of the universe on the grounds of expense and risk of life; however man is still the most compact computer known with the questionable advantage from a publicity angle, of guaranteeing more prestige to those who land him on the moon than to those who land even the most sophisticated robot. It still seemed certain, then, that the two great powers were racing towards a manned landing.

The first manned Gemini mission finally lifted off Pad 19 on 23 March 1965, carrying John Young and 'Gus' Grissom, the Mercury veteran. Grissom had ended his Mercury flight paddling alone in the Atlantic after his capsule, Liberty Bell, sank beneath him. After his fellow astronauts had given him a good deal of ribbing about his enforced seamanship he named his new Gemini capsule, Molly Brown, after the Broadway musical, *The Unsinkable Molly Brown*. Thus the manned Gemini programme began, after lift-off had been achieved, with the laconic retort from Flight Control: 'You're on your way, Molly Brown'. The custom of naming spacecraft ended abruptly when the 1968 crew of Apollo 7 suggested calling their capsule Rub-a-Dub-Dub. Permission was refused.

Molly Brown flew just three orbits in nearly 5 hours, and apart from a minor panic over a faulty oxygen pressure reading the mission went smoothly – in spite of the crumbs from the famous corned beef sandwich.

The effect of weightlessness still preoccupied NASA's medical team, and one objective of the mission was to find out if it had any sterilising action on living organisms. Amongst the scientific experiments they carried out was the fertilisation of sea-urchin eggs.

The landing was more abrupt than either astronaut had expected: they were unprepared for the sharp jolt as the parachutes opened, and, being only loosely strapped to their couches, both crewmen got badly bruised. Grissom was the unluckier – a control knob smashed through his face plate, and then he was sea-sick while awaiting recovery after splashdown.

Gemini 3 was regarded as the last of the developmental flights, and 4 and 5 were to be marathons, collecting medical information, and doing most of the strictly scientific experiments, as well as 15 of a military nature. Extravehicular activity was scheduled for later docking missions, and it was very much a last minute decision which allowed White to parallel Leonov's achievement and go out for a space walk on Gemini 4. The spacesuit and self-propulsion system were only ready 10 days before the launch.

On the morning of 3 June 1965, one anxious hour after the countdown should have ended, Gemini 4 lifted from Pad 19. Just over three hours later the cabin was depressurised, the hatch opened (no heavy airlock was needed as in Voskhod) and Ed White plus a loose glove drifted into space to become the first spaceman to truly control his movement. Unlike Leonov, White relied on oxygen supplied through a long pliable tube which became known as the 'umbilical'. He enjoyed his excursion so much that he stayed out for 21 minutes, considerably longer than originally planned. Afterwards he denied allegations that he might have been suffering from 'space euphoria', a suggested space hazard similar to deep-sea narcosis. He reported no sensation of either speed or disorientation. He manoeuvred in space using a hand held gas gun which proved highly efficient until it prematurely ran out of fuel – another unwanted first for America: the first man in space to run out of gas.

White and McDivitt stayed up for nearly 100 hours, completing 62 orbits. They found alternate watch-keeping, with one crewman asleep while the other worked, an unsatisfactory routine, and both slept at the same time: this practice became standard on all Gemini flights.

On this flight the computer broke down, making an automatically controlled re-entry impossible; however they splashed down only 50 miles short of target, having made a Mercury type of re-entry.

The next Gemini flight was the nearest thing to a failure during the whole manned part of the Gemini programme. On 21 August 1965, after a delay lasting two days, Gordon Cooper and Charles Conrad were launched on G-T 5 to begin the first truly long duration flight. Gemini 5 was the first mission to be equipped with the new oxygen/hydrogen fuel cells (designed for the Apollo project) instead of batteries. They had just begun to test a rendezvous radar in the nose of the spacecraft when the new fuel cells began to give trouble. It was established later that a heating unit had failed, and oxygen pressure began to fall rapidly. The radar test was abandoned, and for a time it looked as though the whole mission would have to be called off. Finally, as the oxygen pressure steadied, it was decided to carry on, concentrating on the medical aspects of the flight. Only when the fuel cell problem eased were they able to do some minor manoeuvres with the spacecraft, conserving power for re-entry.

They stayed in orbit for nearly 191 hours, orbiting the earth 120 times. At the end of the flight an error of information fed from the ground to the on-board computer caused them to splash down over 100 miles from their target but Cooper and Conrad were in good shape, and valuable information had been gained about both astronaut endurance and fuel cells.

The next phase of the Gemini programme was to be the most ambitious ever undertaken in space, a series of rendezvous and docking missions.

Overleaf *Ed White's space walk.*

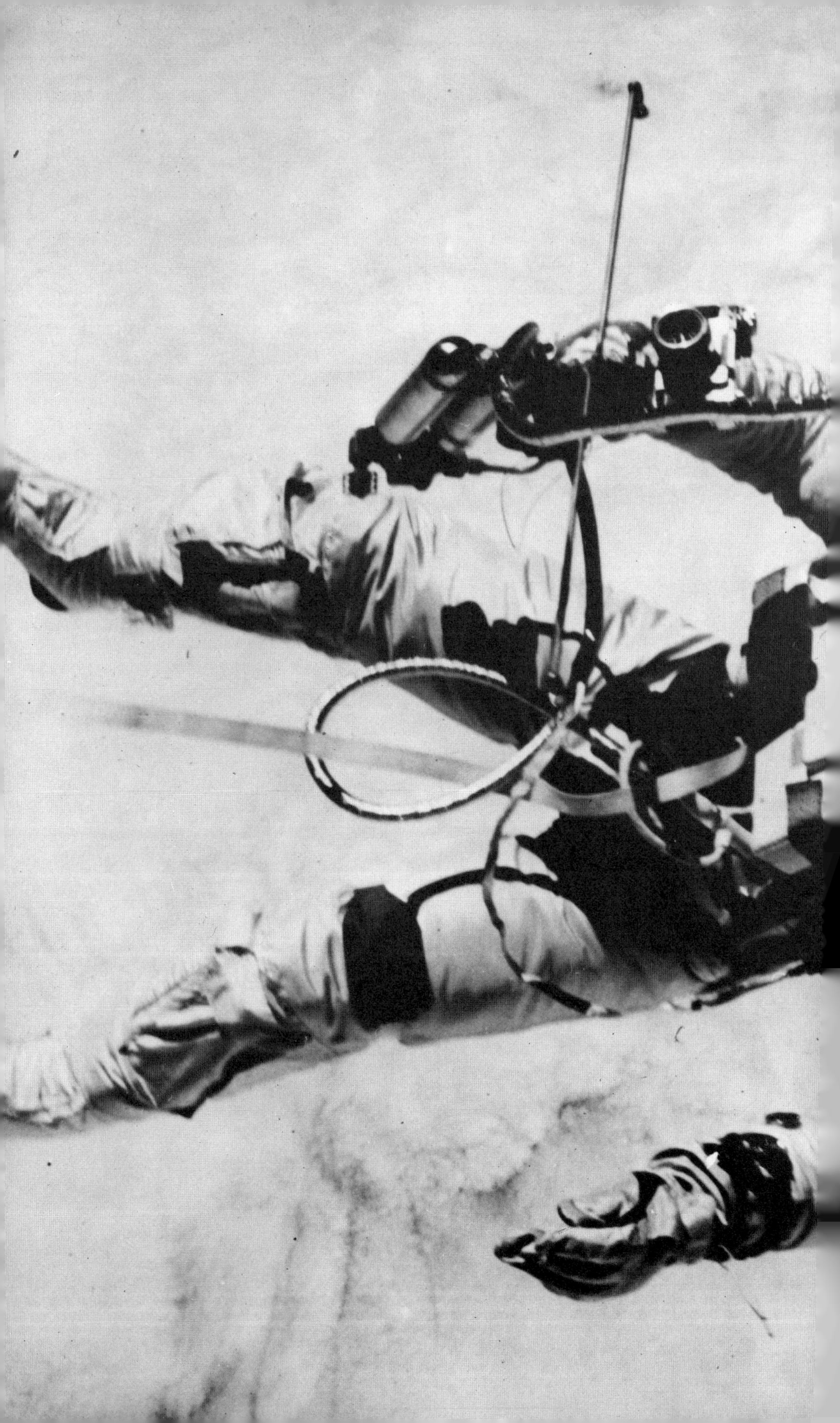

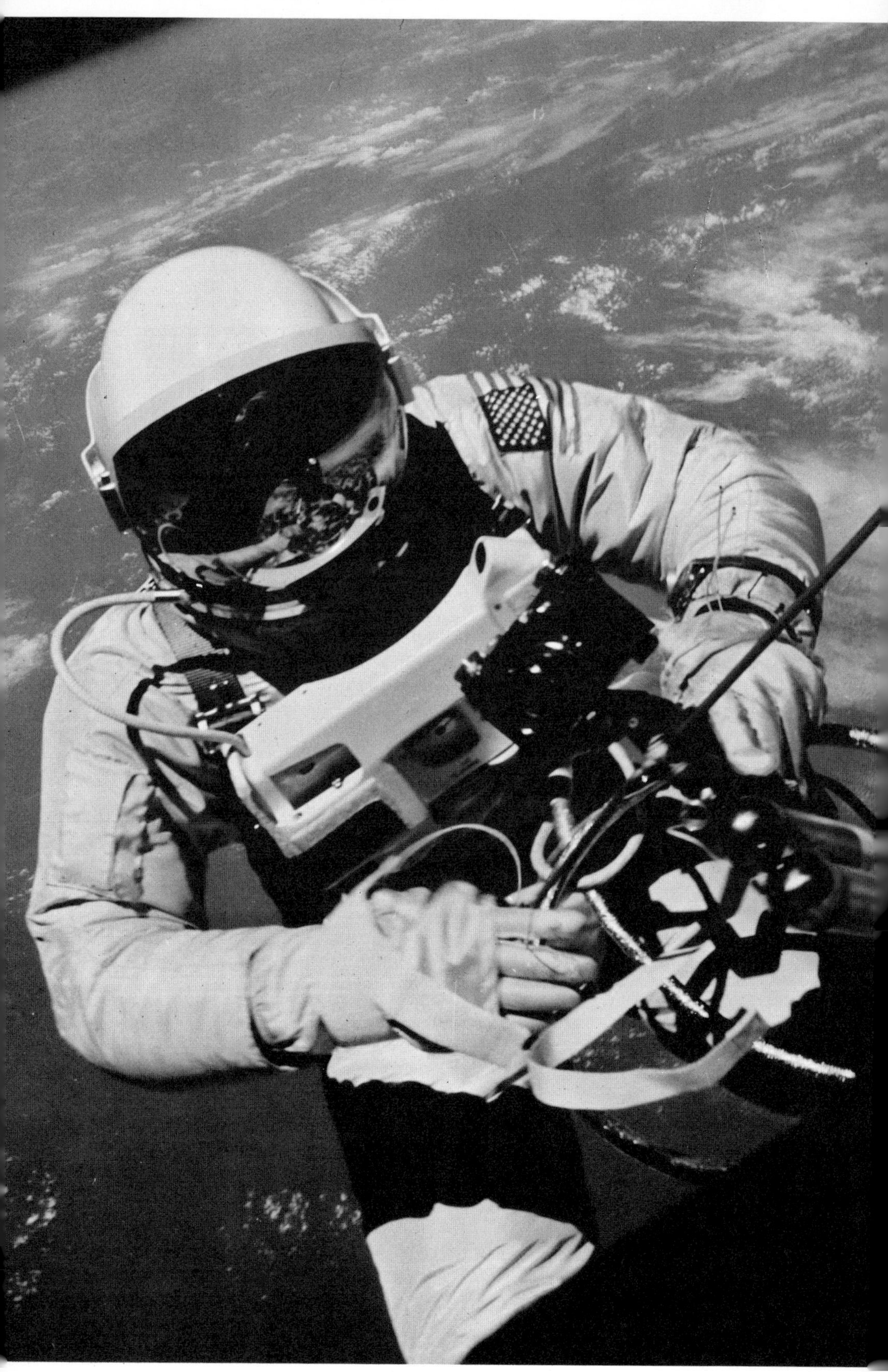

Left *Gemini 4. Ed White floating in space during his third orbit. He was secured to Gemini 4 by a 25-feet umbilical line and a 23-feet tether line, both wrapped together with gold tape to form one cord. Wearing a specially designed space-suit for EVA, he remained outside for a total of 21 minutes.*

Above *A view of the orbiting Gemini 7 spacecraft, taken from Gemini 6 during their historic rendezvous mission. At the moment of this photograph the craft were about 13 feet apart. In Gemini 6 were Schirra and Stafford, and in Gemini 7, Borman and Lovell.*

Above *The first Cosmonauts in Red Square on 19 October 1964, in honour of the Voskhod group space flight. Left to right: Gagarin, Titov, Nikolayev, Popovich, Nikolayeva-Tereshkova, and Komarov.*

Below *Kruschev toasts the bride and groom at the Tereshkova-Nikolayev cosmonaut wedding. Gagarin stands on Kruschev's left.*

Above *An early space meal, eaten by Nikolayev and Popovich. At that time the hazard of crumbs in the spacecraft was presumably not appreciated.*

9 The Robot Explorers Ranger, Lunik, Zond, and Surveyor

With cosmonauts and astronauts orbiting the earth at incredible speeds, many people must have been puzzled why it was not merely a matter of days before whole platoons of spacemen were planting national flags into the surface of the moon. As a target however, the moon, despite the fact that it looks remarkable close in our night sky, demands great accuracy of aim. To escape completely from earth's gravitational field a spacecraft must reach 35,700 feet per second. To escape to the moon the speed must be almost as fast – 35,300 feet per second. At that speed an inaccuracy in velocity of only one foot per second could cause a spacecraft to miss by 125 miles; an error in direction of only one degree would cause a miss of 7,000 miles. Finally in the 1950's there was almost total ignorance about the nature of the target itself, and many experts still believed that the moon's surface was covered with a thick coating of fine dust which would engulf any object foolish enough to attempt a landing. Both the United States and Russia agreed that, before any man could be sent to the moon, robots would have to prove the possibility of such a journey, and both chart and physically test the landing sites.

The earliest series of American moon probes, Pioneer, dated from pre-NASA days, although all but the first attempted launch took place after NASA was formed. The US Department of Defence instigated the Pioneer series as an International Geophysical Year project, but if it ever had any unrevealed military prestige objectives like showing American missile potential, the Russians must have been highly relieved: the early Pioneer series was a colossal flop.

Inter-service rivalry bedevilled the project. On 17 August 1958 the Thor-Able, an Air Force and Navy inspired rocket blew to pieces 77 seconds after lift-off. The Air Force tried again in October with an identical package now called Pioneer 1. The Air Force Thor IRBM first stage worked perfectly, as did the second stage (an adapted version of the Navy's Vanguard rocket). However, the third stage malfunctioned, and a third of the way to the moon it turned round, plunging back towards the Pacific Ocean. Fortunately some usable information about radiation belts and micro-meteorite impacts was sent back before the flight aborted.

By November 1958 NASA itself had taken a firmer control of the series, but the Air Force Pioneer 2 was equally ill-fated and aborted after its third stage failed to fire. In December, using a hybrid rocket with an Army Jupiter IRBM as first stage, surmounted by three solid fuel upper-stages, a 13-lb. gold-plated capsule containing radiation instruments was launched as Pioneer 3 – and failed to reach even the height of Pioneer 1. Then, on 2 January 1959, the Russians launched Lunik 1. It missed the moon by nearly 4,000 miles, but in terms of the space race it was useful for prestige, for it had at least succeeded in escaping from earth's gravity. In their announcement of the flight, the Russians made a big point that Lunik 1 had become the first artificial satellite of the sun. The American climate of jealous frustration was illustrated by one US

Army official, who was quoted as saying: 'What's so great about orbiting the sun? Any time I throw a baseball in the air it's orbiting the sun.'

Only a few weeks later the Americans themselves were indulging in self-congratulation for having launched an artificial sun satellite. Pioneer 4 was ten times further off target than the Russian Lunik 1, but remained the closest thing to a success in the entire Pioneer programme, and transmitted data on radiation and temperature for 3 days. The flight encouraged the US Army to use the considerably more powerful Atlas ICBM in combination with the Able rocket, to launch a much heavier robot spacecraft towards the moon. It weighed 372 lb., and, apart from carrying some complex radiation measuring gear and a magnetometer, it was equipped with a camera scanner, and a data memory to store information for later transmission to earth. Before it could be launched the Russians had chalked up yet another important space first: on 4 October 1959 a multi-stage rocket sent Lunik 3, an unmanned vehicle, *en route* for the moon. It carried a single photographic camera whose pictures could be processed automatically on board, and then radioed back to earth. Its orbit was devised to loop the moon, enabling the probe to take the first photographs of the back of the moon never previously seen from earth. To do so the probe incorporated some remarkably sophisticated control and guidance systems: a gyro system, coupled to a photo electric cell, manoeuvred it so that its camera was always looking at the moon with the sun behind. It took pictures for a whole year, enabling the Russians to compile and publish a detailed atlas of the far side of the moon in 1960. The individual photographs were of poor quality, and somewhat spoilt by interference during their transmission, but, by using the latest techniques of photo interpretation they were considerably improved and again impressed the world with Russia's technological mastery. Both Tsiolkovsky and Jules Verne were appropriately honoured by having craters named after them and inevitably the name Lenin became another lunar landmark.

On 26 November 1959 the Americans tried to match the Russian achievement. Atlas-Able 4 was launched, and within 45 seconds its nose cone was torn off by vibration. Nearly a year elapsed before the launch of Atlas-Able 5A, but this time the flame died after the second stage had achieved separation, and the rocket plunged into the sea.

NASA had now devised the Ranger and Surveyor programmes, and pressure was mounting to end the Pioneer series. It was decided to have one more try and on 15 December 1960, 70 seconds after lift-off the Atlas-Able 5B closed the first Pioneer chapter with a bang by totally disintegrating.

Originally NASA had planned three different programmes for robot investigation of the moon: Ranger was to be a moon impact craft, followed by Surveyor, which would study both soft-landing and lunar orbiting, and Prospector was to be a mobile lunar laboratory research programme. During 1961 the Prospector programme was cancelled, and the two parts of the Surveyor programme were rescheduled to run concurrently. NASA had to be certain that robot studies would be complete before man himself was ready to go to the moon.

The Pioneer fiasco had shown that American rocketry was not yet ready for a direct moon shot. Because of this, the restartable Agena rocket was chosen as the second stage for Ranger, to boost the spacecraft into a parking orbit around the earth, before reignition for the flight to the moon.

On Rangers 1 and 2 the Agena failed to reignite, and both spacecraft were left in a decaying earth orbit from which they eventually descended and burnt up in the atmosphere. These first two missions were intended as proving flights to ensure the success of the next three; another proving flight would have delayed the programme by an additional six months, so on 26 January 1962 Ranger 3 was launched. It was equipped with a gamma-ray spectrometer to transmit readings throughout the flight, and television cameras designed to take a picture every 13 seconds from 2,500 miles from the moon down to 8 seconds before impact. It also carried a seismometer to measure the tiny tremors predicted on the moon surface (caused, for example, by meteorites). The seismometer was sensitive enough to register the impact of a 5-lb. meteorite on the other side of the moon. Elaborate precautions were taken to ensure the safe landing of this delicate instrument. At 14 miles above the moon it was to be ejected from Ranger (which would impact at 6,000 mph), and slowed by retro-rockets to an impact speed of 200 mph; the primary shock would be absorbed by its 7½-inch thick balsa wood casing. The instrument was given additional protection by being suspended in a thick oil which filled the capsule. After landing two bullets would be fired through the metal shell, permitting the oil to escape, and the seismometer to be gently left high and dry, sensing the slightest movement beneath it. A battery would enable it to transmit its findings back to earth for about 30 days. After a perfect insertion into its parking orbit around the earth, Ranger 3 was aimed at the moon. It missed by 22,862 miles.

In April Ranger 4 was sent on a similar mission, and hit its target – but the wrong side of it: both the telemetry transmission and the main power supply failed on this flight, and no information was relayed back to earth. Six months later Ranger 5 again missed the moon, this time by 450 miles, and because of a total power supply failure shortly after its launch, it again sent back no data.

Concern about this wasteful series of failures was now so great that the launches of the next four Ranger missions were postponed, and the programme re-evaluated: many suggested totally abandoning Ranger, and concentrating on Surveyor. Finally NASA decided to scrap the seismometer part of the flight and concentrate on photographing possible lunar landing sites.

In January 1964 the launch and trajectory of Ranger 6 were perfect, but the cameras accidentally switched on in mid-flight, and by the time Ranger approached the moon there was insufficient battery power to transmit anything from the six television cameras. $28,000,000 had been spent in obtaining 4,000 pictures of empty space which nobody had even realised were being transmitted – until the signal was so low as to be useless. Ranger 7 was delayed while the TV system was modified, but, after three postponements, it lifted into the intended parking orbit on 28 July. The Agena rocket fired correctly and Ranger 7 began its journey towards the moon. A 50-second burst on the hydrazine engine corrected the course to steer for an impact in the Sea of Clouds. At 6.07 Pacific Daylight time on 31 July 1964 the first of 4,316 photographs was received at Goldstone. For 17 minutes the cameras functioned perfectly until Ranger 7 smashed itself to pieces (as intended) only six miles from its precise target. The photographs obtained were 2,000 times more detailed than any seen before; a possible lunar landing site was identified, and Ranger was suddenly a success. In April and March of 1965 Rangers 8 and 9 added another 12,951 superb photographs to the NASA lunar scrapbook. However the Ranger programme had not fulfilled many of its original objectives, and the dust dilemma was still unresolved. To study potential landing sites the programme could be called the most expensive preliminary 'recce' in history.

After the first three, brilliantly successful lunar probes in the Lunik series, all was now surprisingly quiet on the Russian front: their attempts to land robot spacecraft had ceased abruptly for five years with the single exception of Lunik 4, which passed within 5,400 miles of the moon in April 1963 on a minimum energy lunar trajectory. Then, seven weeks after Ranger 9, the Russians suddenly began an intensive series of lunar probe launches.

The official Russian account says the series went exactly as planned, with four proving flights, before a successful attempt at soft-landing with Lunik 9. Considering the American proportion of admitted failures it is only human to suspect however that the Russians attempted to soft-land four times before succeeding. The launch weight of the highly advanced Lunik spacecraft was about 1½ tons, requiring a very powerful launch vehicle. During a gap in the programme an automatic spacecraft from the Zond series (Zond 1 had been sent to Venus on 2 April 1964, and Zond 2 to Mars on 30 November 1964) was sent past the moon to photograph the part of the reverse side which had been in darkness during the Lunik 3 mission. In October and December 1965 two more Luniks (7 and 8) crashed into the moon; it is supposed that they again failed to soft-land, for Lunik 7 hit the Ocean of Storms at only 45 mph. Unlike the Ranger spacecraft the Luniks were not equipped with cameras to take pre-impact pictures, so the missions were presumably regarded privately as total failures. Lunik 9, however, was a remarkable success. It was launched on 31 January 1966, and at 7.45 p.m. on 3 February it settled gently on to the lunar surface, to make the first soft-landing on the moon. It was, so far, the heaviest spacecraft in the series, weighing

3,490 lb.; 50 miles above the moon a 220-lb. spherical landing capsule was detached to land in the Ocean of Storms between the craters known as Galileo and Cavalieri.

Immediately after landing, four pointed 'petals' forming the top of the sphere opened to expose the camera lens, and act as radio aerial dishes. Three panoramic views of the landscape were transmitted, each made up of 8 photographs taken while the camera revolved at an angle of 30 degrees from the ground: an international sensation resulted, for the picture-transmission was intercepted and recorded by the British radio telescope at Jodrell Bank. Sir Bernard Lovell enlisted the help of a national newspaper to decode the signals by trial and error on an ordinary wire photo receiver. The Russians were confronted with front-page pictures from their own spacecraft before any but the space centre élite even knew that Lunik 9 was taking photographs. The photographs provided data about the condition of the lunar surface, and proved beyond much doubt that it could support a landing.

The next three Lunik probes (10, 11 and 12) were all moon orbiting spacecraft, studying magnetic and gravitational fields, computing the density of micro-meteorites, and analysing the composition of the lunar soil from a height of several hundred miles, supporting the theory that it is of basaltic type.

In addition Lunik 12, launched in October 1966, carried cameras which photographed a belt around the moon in very considerable detail, and seemed to confirm suspicions that the Russians were selecting landing sites for manned missions.

The series ended with a soft landing by Lunik 13 on Christmas Eve, 1966. Apart from sending back yet another excellent set of photographs, it measured the density and firmness of the lunar soil, and removed doubts about the weight-bearing properties of the moon's surface. The Lunik series was a superb demonstration of Soviet skill in space technology, and because the Academy of Science was prepared to divulge many of the findings, made a major contribution to international knowledge.

Early in 1966, after all the delays and re-thinks caused by the Ranger failures, the first Surveyor was at last ready to be sent on a direct lunar trajectory aboard its Atlas-Centaur rocket. NASA decided to make the flight open to world television, and on 30 May 1966 100,000,000 people watched the Atlas-Centaur rise from Cape Kennedy. Sixty-four hours later they heard the cheer in the control room in Pasadena as the calm announcement was made 'We have touched down. . . .' A few minutes later the first blurred picture came back, showing the landing leg. Then, gradually as the first series of pictures were processed and put together jigsaw fashion, the moonscape was revealed as a spectacular rock-strewn panorama. Altogether Surveyor 1 transmitted 11,150 pictures, including colour composites.

The first Lunar Orbiter was circling the moon two and a half months later. Instead of the somewhat crude television scanners of Surveyor, this space robot carried its own portable film processing laboratory. An ingenious camera system exposed pictures on 70 mm film: two shots were taken at a time, one wide angle, and one a high resolution close-up of the centre of the view. Every time a pair of shots was taken, the exposed film was automatically wound into a developing 'bath', producing negatives which were then scanned electronically for television transmission back to earth. Not only was a high resolution photographic map of the moon transmitted, but also the first views of planet earth seen from its satellite.

In April 1967, in spite of bouncing badly on landing, Surveyor 3 successfully used its soil sampler for the first time, digging four trenches, and making load-bearing and penetration tests. Then, in November Surveyor 6 made the first take-off from the moon, rising 12 feet from the surface and landing a planned 8 feet away. From the two positions it transmitted stereoscopic pairs of photographs for comparison.

The programme ended with Surveyor 7 in January 1968. There had been only two failures, and the series had amassed over 87,000 photographs. Surveyors 5, 6 and 7 showed that the moon surface was rather like the volcanic lava on earth, composed of silicon – 19%, aluminium – 7%, oxygen – 57% in oxide form, calcium – 6%, iron – 4%, magnesium – 3%, carbon and sodium – each 2%.

The eventual success of the robots made many ponder the wisdom of the proposed manned missions – especially in the light of the Apollo and Komarov tragedies. Unlike man, robots are expendable, and because of this cheaper to launch, as they need not return. Moreover man requires food, water and oxygen to keep him going, and can only work efficiently within a very limited range of temperature and pressure. Robots can be less heavy, run on sunlight power, work in a vacuum, and tolerate a wide range of temperatures.

As late as November 1968, after the flights of the unmanned Zonds 5 and 6, to the moon, eminent scientists like Sir Bernard Lovell were confidently predicting that Russia's main interest lay in exploring the moon with robots. He went on to say that they were already capable of collecting a rock sample and returning it to earth – 'whether it was dug up by man or machine does not matter'.

With the success of Apollo 7, and America's seeming lead in manned flight developments of that time, it was thought that Soviet interest now almost wholly focused on unmanned exploration. Zond 5, launched on 15 September 1968, seemed at first to confirm this: it chalked up yet another space-first for the USSR when it went round the moon and returned to a landing on earth. However speculation grew after a Russian voice was heard from it reciting instrument readings; later it was revealed that this was a recording, but the craft was admitted to contain two turtles, and some wine flies and meal worms: clearly the Soviet scientists were still very much interested in the problems of life in outer space. However, while all the aerospace Press were at Cape Kennedy and Houston for the flight of Apollo 7, a leading Soviet spokesman,

Leonid Sedov, held an almost unnoticed Press conference. He stated that a manned flight to the moon was 'not now an item on the agenda'. Confusion increased, but six days after the return and pinpoint landing of Zond 6 on 17 November 1968, an article published in the Soviet Press revealed that the purpose of Zonds 4, 5 and 6 was to 'perfect the automatic functioning of a manned spaceship that will be sent to the moon, and to check the functioning of the station's on-board systems in conditions of a flight in the direction earth-moon-earth'. The article went on to say that 'biological specimens' had also been carried on Zond 6, to continue the Zond 5 investigation of radiation hazards. On 25 November *Izvestia* revealed that, despite the high solar activity recorded during the flights, the radiation dose was about one hundredth of the permissible level for humans. Academician Boris Petrov also admitted that the Russians were using a different re-entry technique from the Americans. Zond 6 carried out a kind of double bounce or 'skip' on the outer fringes of atmosphere, approximately halving its speed of 6·8 miles a second by dipping before making a final descent, which was eventually checked by parachutes.

The Russians use this method of re-entry to reduce the heat and 'g' loads on the capsule caused by a straight re-entry; also it saves on fuel, since retro-rockets are needed less in the process of slowing the space vehicle into a re-entry path. While the American manned space programme was making the headlines at the end of 1968, the Russians were pressing ahead with other unmanned research projects; on the eve of the Zond 6 launch the world's biggest automatic space station Proton 4 was put into earth orbit. It weighed 17 tons, more than three-quarters of which was scientific apparatus. Its declared objective was to continue the work of Protons 1, 2 and 3 in studying cosmic rays of high and super-high energies, and their interaction with atomic nuclei. Almost certainly this research programme is part of some very advanced project which could involve the next generation of space propellants—the nuclear devices. By the time the moon has been fully explored by man, our interest may have already switched to the interplanetary possibilities offered by the findings that such unmanned satellites are transmitting today.

bove One of the three
anding pads of Surveyor 1
fter it made America's first
oft landing on the moon on
June 1966. The lunar soil
the vicinity of the craft
as revealed to be granular
nature.

Above *The surface sampler of Surveyor 3, shown in a multi-exposure photograph after having dug several trenches in the moon for onboard analysis. The results indicated that the surface was probably basaltic in origin.*

bove *A scale model of the
Lunar Orbiter spacecraft,
howing how it approaches
o less than 30 miles from the
noon, taking high resolution
ictures of the lunar surface.
t discerned objects as small
s a yard square.*

Above *Assisted by frogmen, Astronauts Walter Schirra and Thomas Stafford practise the techniques of leaving the Gemini capsule after splashdown.*

Right *The moon, showing some of the landing sites of the early unmanned exploration vehicles launched by the US and the USSR.*

Overleaf *The moon, showing the planned photographic sites for Lunar Orbiter 3, an essential preliminary to establishing the actual landing sites for Apollo.*

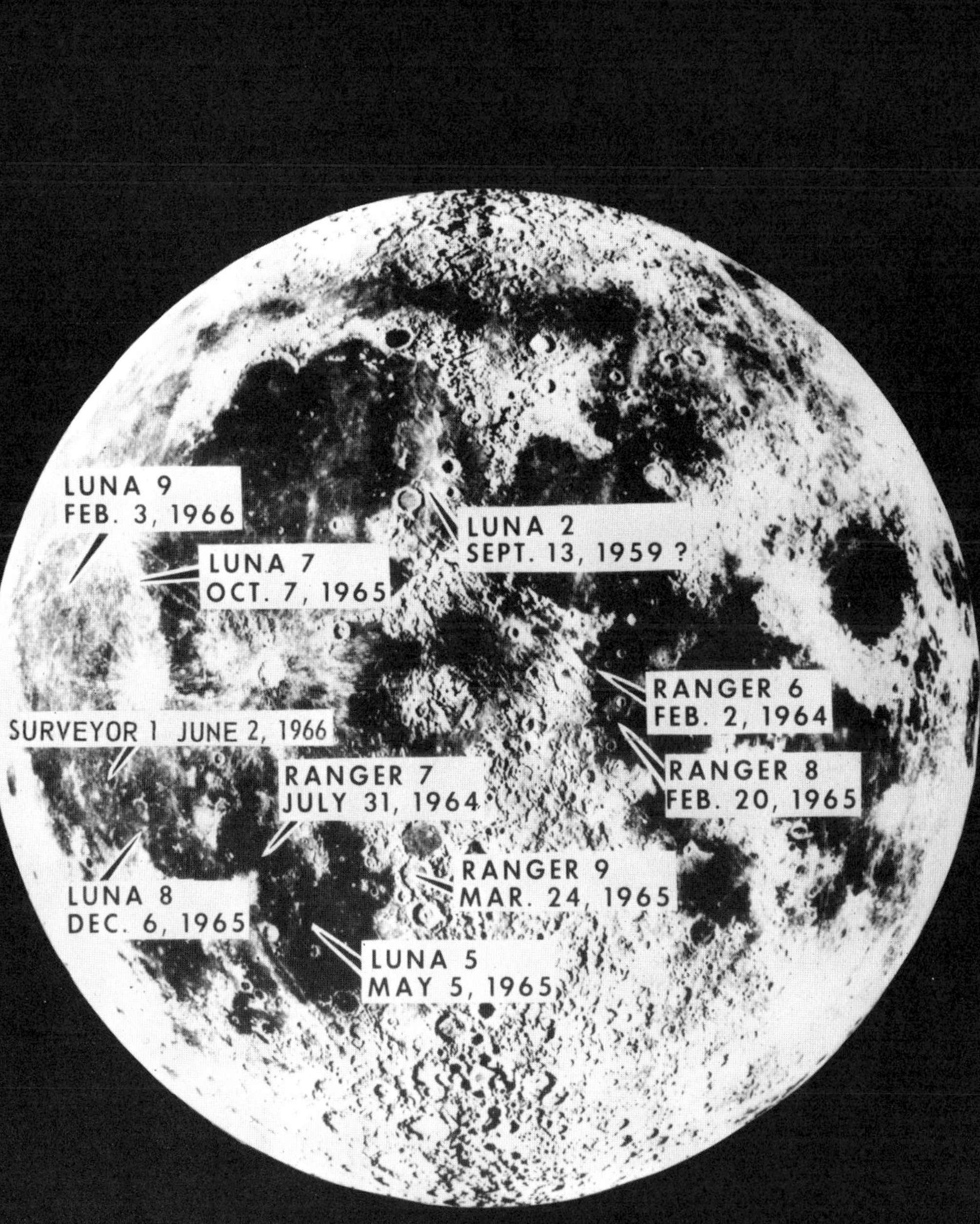
LUNA 9
FEB. 3, 1966
LUNA 7
OCT. 7, 1965
LUNA 2
SEPT. 13, 1959 ?
RANGER 6
FEB. 2, 1964
SURVEYOR 1 JUNE 2, 1966
RANGER 8
FEB. 20, 1965
RANGER 7
JULY 31, 1964
RANGER 9
MAR. 24, 1965
LUNA 8
DEC. 6, 1965
LUNA 5
MAY 5, 1965

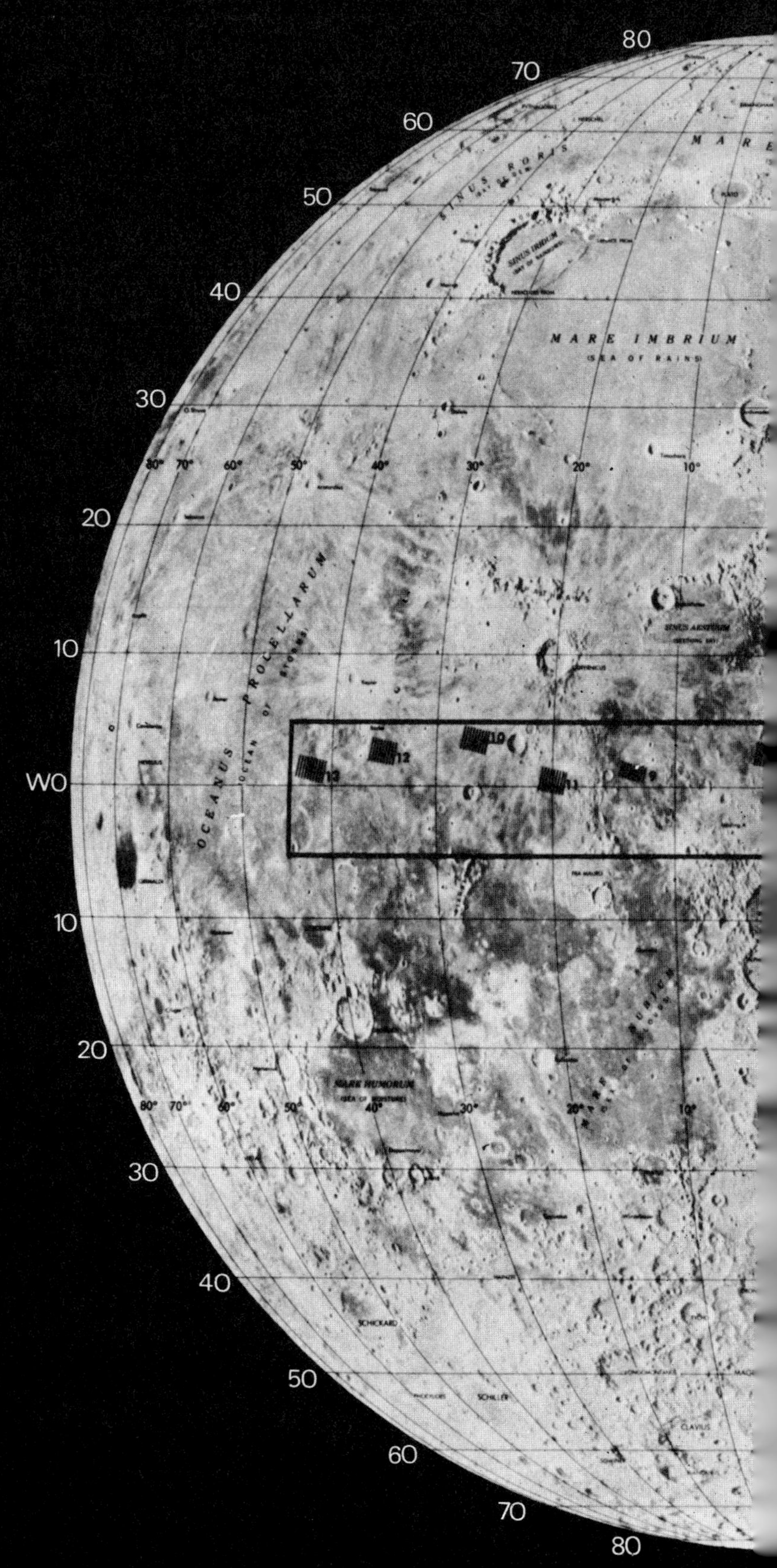

80
70
60
50
40
30
20
10
W0
10
20
30
40
50
60
70
80
MARE IMBRIUM
(SEA OF RAINS)
OCEANUS PROCELLARUM
MARE HUMORUM
SCHICKARD
SCHILLER
CLAVIUS

MARE SERENITATIS
(SEA OF SERENITY)
LACUS SOMNIORUM
TRANQUILLITATIS
MARE NECTARIS
FECUNDITATIS
10°
20°
30°
40°
70°
50°
60°
OE

Above *The moon's hidden side. This photograph taken by Lunar Orbiter 3 has been described by US space officials as one of the best portraits ever made of the hidden side of the moon. The black area near the centre is the 150-mile wide crater, in the middle of which lies the hill which the Russians have named after their pioneer Tsiolkovsky.*

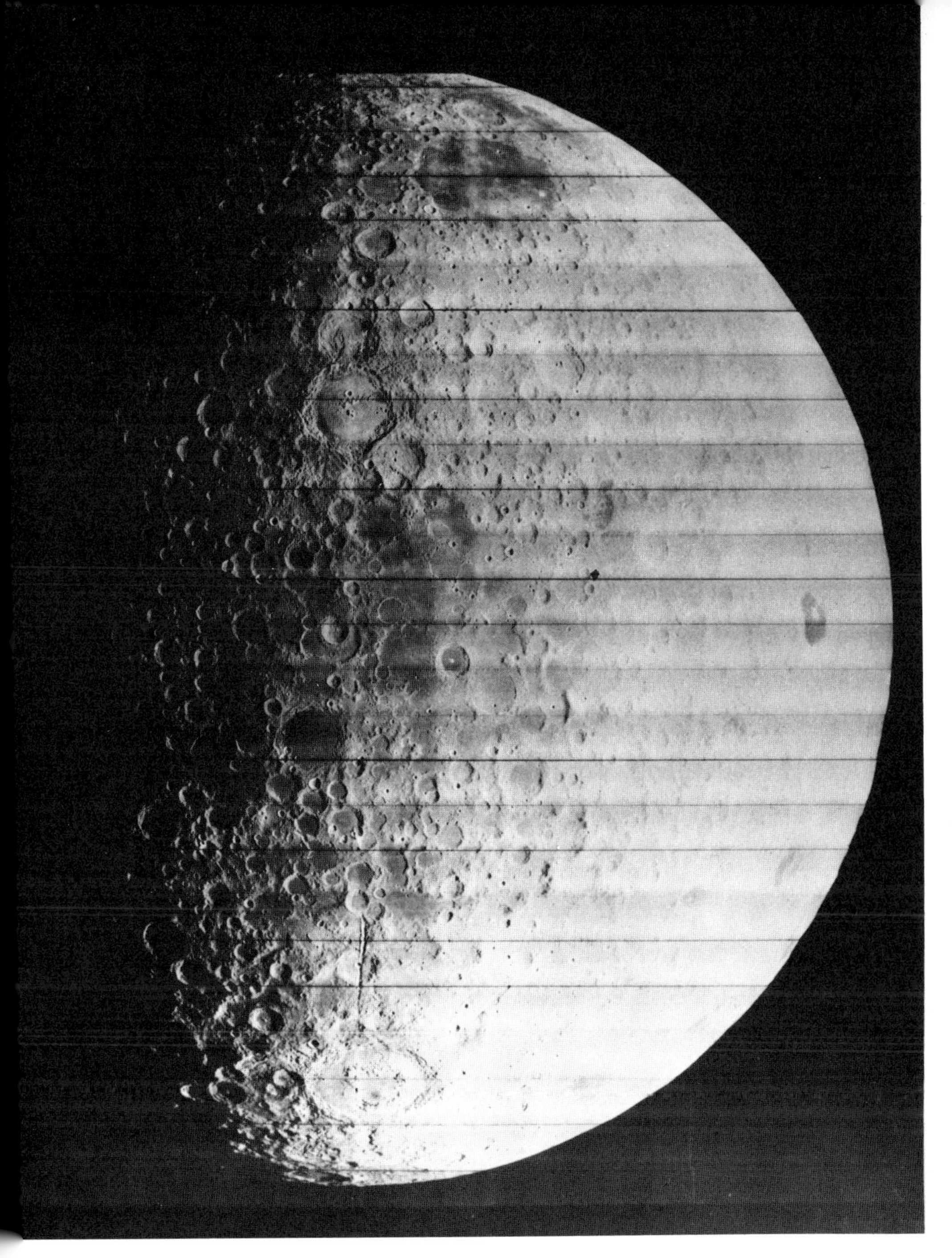

bove *A fault line on the*
ıoon's hidden side shows as
crack 150 miles long, and
miles wide in some places.
ı Lunar Orbiter 4
hotograph.

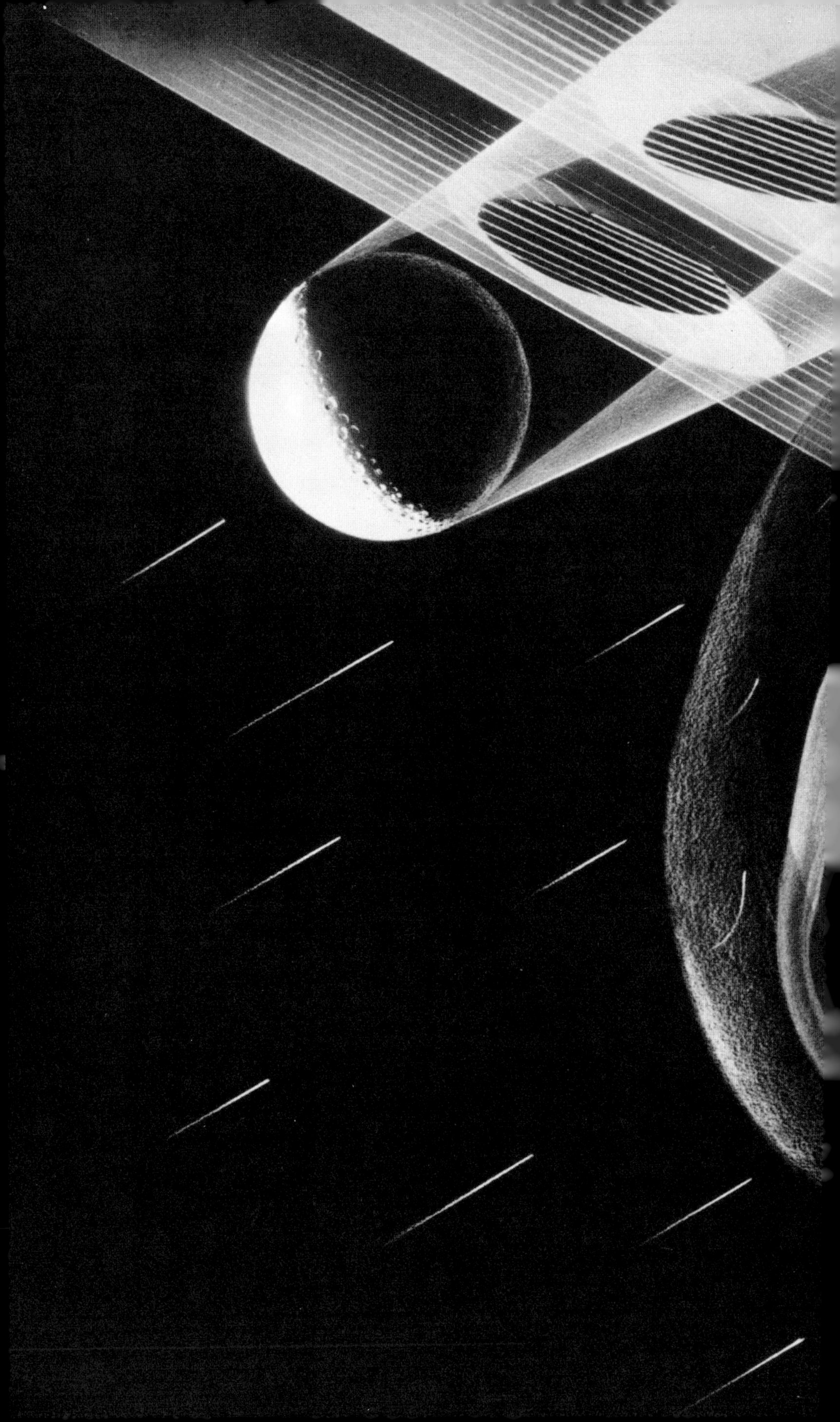

Space Marathons Americans and Russians spend days in space

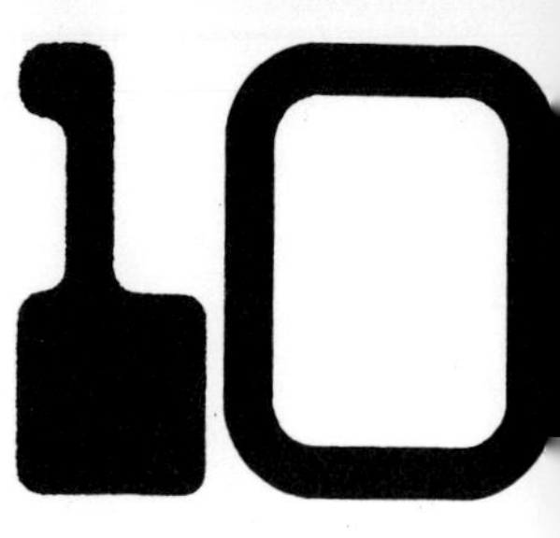

Pages 130–1 *Findings about the moon and its environmen made by the Explorer 15 satellite, showing that the moon has almost no magneti field, and hence, no magnetosphere envelope like the earth. This artist's impression shows the solar wind (white dashes) streamin at a million miles an hour straight to the moon, while the shock front preceding th earth deflects the wind. In th lunar shadow is a wind void and the interplanetary magnetic field lines (diagona solid lines) are distorted. Scientists say this region extends more than 100,000 miles behind the moon. Earth's magnetosphere (closed region behind the shock front), formed by the magnetic field, is compresse on the bright side, and distended on the dark side. This complex envelope helps shield the earth from potentially harmful radiati The tail of earth's magneto-sphere extends several millio miles into space.*

On 25 October 1965, Wally Schirra and Tom Stafford were sitting in the capsule of Gemini 6 carrying out the pre-flight checks ready to lift off and attempt the first rendezvous and docking operations of space history. Their target was to be the upper stage of an Atlas-Agena. To avoid the ambitious exercise of attempting this first docking with another manned spacecraft, the Agena had been chosen as a passive target, to be launched one orbit ahead of the Gemini capsule. It had the advantage of being powered by a hypergolic fuel system, which is easily restartable in space because its fuel (di-methyl hydrazine and an oxidiser) ignites on contact, and can therefore be controlled almost as easily as turning a tap on and off. Once in orbit the Agena upper-stage, equipped with a docking collar which would latch to the nose of Gemini, and a set of external instruments like an illuminated car dashboard, would become a flying fuel tank to propel the capsule, once docking had been accomplished.

The countdown went smoothly, and 95 minutes before Gemini 6 was due to lift off, the time lapse needed for the target to complete one orbit, the Atlas-Agena rose into the sky. Six minutes later, after it had successfully separated from its booster, the control team watching the radar screens of the Atlantic tracking network saw its trace disappear. Not a single nut or bolt of it was ever seen again. Something had gone wrong with its normally dependable motors, and the whole assembly had blasted to smithereens. With less than an hour and a half to go before ignition of Gemini 6 the question was whether another vehicle could be substituted as a target, or a less ambitious programme be resolved for Schirra and Stafford. Reluctantly the Flight Director told the astronauts that it was 'No Go'.

To prepare another target vehicle quickly was out of the question, but NASA wanted a space rendezvous by the end of the year. One more long duration mission was due in the Gemini programme, G-T 7, due to lift off in early December. In a mood of supreme confidence the idea was put forward of using G-T 7 as the rendezvous target for the postponed G-T 6. At Cape Kennedy there was only one launch pad suitable for the Gemini-Titan blast-offs, Pad 19. To marry the two missions would mean sending Borman and Lovell up in Gemini 7, and then preparing Pad 19 for a second launch in record time. After a lift-off scorching, the pad has to be extensively repaired, and the next vehicle, after being placed in the firing position would need a vast series of pre-flight checks before it could be launched. In the case of Gemini-Titan about 400,000 checks were standard.

The ground staff at the Cape were instructed to find out if all the work could be compressed into ten days: Gemini 7 was scheduled for a fourteen-day flight, so, if Gemini 6 could be made ready in time, there was a chance of achieving a rendezvous. The ground staff warned that it would only be possible if some of the normal pre-flight checks were ignored.

Considering all the dangers involved, it seems amazing today that NASA decided to take the risk; but just after 2.30 p.m. on

4 December 1965, G-T 7 took off to lift Borman and Lovell into the longest space flight ever attempted. Pad 19 had scarcely cooled before engineers were beginning to prepare it for the second launch. Twenty-four hours later the massive Titan rocket that had stood ready for launching just 42 days earlier was again ponderously set in position; then began probably the most superficial pre-flight check routine in the history of manned space flight. It has been alleged that only a quarter of the normal checks were properly carried out. Eight days later Schirra and Stafford eased into their couches for the second time, and awaited blast-off. At 9.54.06 a.m. on 12 December the Titan engines blazed into life. One second later they shut down: one of the rocket's umbilical cords had dropped off prematurely, and the Malfunction Detection System had automatically switched off the engines. Schirra now had to pull the ring between his knees to eject both himself and Stafford: in that instant it looked as if the whole operation would have to be called off until the following year when the preparations could be done properly.

However confidence was high, and impatience higher: in an almost swashbuckling mood, knowing full well that the MDS had detected a fault which could perhaps cause an immediate explosion, and knowing also that the only hope of getting G-T 6 off the pad in time to meet Borman and Lovell in space, was to stay in the capsule, Schirra decided not to pull the ring.

In a remarkably short time the engineers repaired the fault that had led to the abort, and repaired another which they had discovered. Gemini 7 still had six days left to act as target. Three days later Schirra and Stafford were once again in countdown. At 8.38 a.m. they went up to join their colleagues in space. By 2.30 that afternoon Gemini 6 and 7 were flying in formation. Borman was heard by ground control to drawl over his intercom: 'What kept you?'

The computer controlled orbit manoeuvres were even more precise than had been expected. Step by step in a series of half circles, Gemini 6 approached Gemini 7 until the two spacecraft were only ten inches apart. Schirra found that, within the limits of his fuel supply, he was able to manoeuvre his capsule around Gemini 7 with extreme accuracy. Although Tom Stafford, operating the onboard computer, described himself as 'busier than a one-armed paperhanger', everyone agreed that the problems of rendezvous seemed to be solved. The next stage, docking, would have to wait: it was feared that there might be a catastrophic discharge of static electricity if the two craft actually touched, so after 5 hours 18 minutes (about 4 orbits), Schirra and Stafford waved goodbye to their companions and burned up into a higher orbit, thereby dropping behind by about 30 miles. Then, until they remembered the time of year, there was a mild panic at Mission Control, Houston. Schirra reported a visual sighting of an unidentified flying object in a parallel orbit, powered by a team of reindeer. After two choruses of *Jingle Bells* on a harmonica, Gemini 6 splashed down under full pilot control on 16 December.

Two days later Borman and Lovell landed, after a space marathon of 330 hours, having completed the most intensive medical research programme ever set for two astronauts, providing more information on the biology of space flight than any other mission. The 14-day duration of their flight had special significance for the medical experts: previously they had been worried by the effect of weightlessness on the cardiovascular system of the astronauts. During weightlessness the heart has less work to do, pumping blood around the body: on long duration flights it gets 'lazy', and when astronauts landed it was found that they suffered from heart strain in direct proportion to the time they had spent in space. Research seemed to indicate that, after a flight of 14 days, the strain on the heart could reach serious proportions. Fortunately for Borman and Lovell the 14-day 'heart barrier' proved to be a myth and they recovered very quickly.

The doctors were also preoccupied about calcium mobilisation: the calcium in the bones tends to dissolve into the bloodstream after prolonged periods of physical inactivity, and if this process is allowed to continue the dissolved calcium accumulates in the kidney, leading to kidney stones. The astronauts had very little space for exercise – as Borman said: 'people can project the problem if they imagine living two weeks in the front seat of a Volkswagen'. Previous tests on calcium mobilistion had shown that the danger increased as the flights got longer, but Borman and Lovell were able to prove that there was no cause for serious concern. They had two other major medical investigations to carry out, on their depth of sleep, and decrease in blood volume. They were wired up with an electrocardiogram, a phono-cardiogram, four electro-encephalograph sensors to measure depth of sleep, and a blood pressure gauge which they had to pump up periodically when instructed by the ground staff. A bio-medical tape recorder stored all the information for later study. Borman slept badly during the long flight, and commented later, 'I don't know why I had difficulty in sleeping. Jim Lovell slept very well. I was envious many times when I'd look over and see him snoring away. I think that perhaps the lack of space and the fact that we weren't able to stretch out at all led to it . . .'

Officially it was explained as 'Command Pilot Syndrome' – the urge to keep waking up to check the instruments. On food Borman said: 'The only trouble we had . . . was that I guess the people didn't realise we were going to operate on a regular day, and try to eat breakfast, lunch and dinner. The meals weren't prepared that way, and several times we had shrimp cocktail and peas for breakfast . . .'

Urine is normally disposed of almost immediately (causing a spectacular display of multi-coloured light as it crystallises and floats away into space), but on this flight, as part of the calcium mobilisation experiment, 75 cc of each urination was stored for later analysis. As Borman said: '. . . one of the problems we had was finding how to store . . . the food, the body waste, and the natural accumulation of waste paper and so on. . . . We used every spare inch of volume in the spacecraft. We stored waste food behind the seats, on top of the seats, under the seats, and around the seats.'

The next Gemini flight, carrying Armstrong, the first civilian in space, and Scott, was planned to achieve the docking which Gemini 6 had been unable to accomplish. On 16 March 1966 the Agena target was put into a 185-mile circular orbit. Two orbits later Gemini 8 slid in behind, to begin the complex flight adjustments which would bring the target progressively closer. Six hours after launch Armstrong and Scott were close enough to see the Agena's instrument display. After a final check Armstrong gave the command, and the United States had, at last, achieved a truly significant space-first. The Gemini capsule locked itself neatly to the Agena, and another space myth was exploded: no problems were encountered with static electricity.

Using the attitude control jets they succeeded in turning the linked vehicles through 90 degrees. Then, 27 minutes after docking, Armstrong and Scott found themselves in a real emergency equal to the worst they had ever faced back in the safety of the flight simulator at Houston.

The yaw thruster, one of Gemini's attitude control motors, had jammed in the 'full on' position. For some time the astronauts did not realise this. The Gemini-Agena combination began to spin wildly. The spin built up until the capsule was turning a complete revolution once a second, creating its own artificial gravity inside, and making it difficult even to think. With remarkable calm Armstrong set about discovering the fault. His first thought was that the Agena had malfunctioned, but the instruments showed everything as normal. He knew they must undock immediately, but it would have been suicidal to attempt that while the capsule was spinning: they would be thrown off at a random angle to end up in a completely different orbit. Ground control were powerless to help. By lightning calculation he worked out how to use the other two groups of attitude control jets to partially correct the spin. He battled with the tumbling spacecraft until it was sufficiently stable to dare risk disengaging it from the Agena. With commendable presence of mind he had already reset the Agena's command and communications system so that it would still be of some use after separation. The fault was corrected, and the panic over, but the Flight Director insisted on terminating the mission immediately: without waiting for Gemini 8 to complete her orbit, he ordered her down into a secondary landing area in the Pacific Ocean near Okinawa, and they splashed down just three miles from the arranged point.

Both astronauts were safe, but very seasick. Scott was disappointed: he had been scheduled for a long EVA. The public was disturbed – NASA did not conceal from the Press just how close Gemini 8 had come to disaster.

Gemini 9, the next mission, seemed ill-fated from the start. Originally Elliott See and Charles Bassett were to fly it, but on the way to St Louis for special training they were killed in a plane crash. On 17 May the Target Agena for Gemini 9 took off from Cape Kennedy, and like the Gemini 6 target, disappeared off the screens a few minutes later. Fortunately a standby target had already been prepared: it was a purely static type of satellite, incapable of propelling itself in space, but carrying the standard Agena type of docking collar. On 1 June it achieved a good circular orbit. Ninety-six minutes after its lift-off, Gemini 9 should have followed it – however Stafford, already the victim of the Gemini 6 launch failures, was soon walking back to his quarters yet again: an electronic fault had cancelled take-off.

At last, two days later, Gemini 9 with Stafford and Cernan got off the ground, only to find the biggest disappointment of all waiting for it in space. The protective nose cone of the target, which was supposed to spring apart and be jettisoned once in orbit, had only half opened and was still wedged to the front of the target vehicle. As Stafford put it, Gemini 9 might just as well have tried to dock with an angry alligator: he suggested to ground control that he be allowed to try and nudge the nose cone off with the Gemini capsule, but permission was refused.

Another alarming aspect of this flight was the space walk. Eugene Cernan was required to float to the back of the spacecraft, and unship the Astronaut Manoeuvring Unit from the adapter section. Having strapped himself in, he was to use its propulsion to help him with a number of symbolic maintenance tasks around the outside of the capsule.

Even before he was due to go and collect the AMU he began to sweat profusely and his visor misted over, making it impossible to see what he was doing. It was beyond his capabilities to put the AMU on to his back. He took a rest, but the fogging remained. Bravely Cernan tried to complete his task but the whole thing was clearly too much for him. He was brought in exhausted, the EVA was abandoned, and the perplexed Mission Control team began to try to puzzle out what had gone wrong.

From that flight onwards the Gemini programme went almost without a hitch. On 18 July, Young and Collins were launched aboard Gemini 10 to complete a flawless rendezvous and docking with an Agena target. Having successfully restarted the Agena propulsion system they put themselves into an elliptical orbit with a record apogee of 476 miles. Then, after returning into a circular orbit they undocked from Agena 10 and successfully located, rendezvoused and docked with the Agena 8 which was still in orbit.

On Gemini 11, although the docking manoeuvres were once more highly successful, the EVA carried out by Richard Gordon again proved far more strenuous than had been predicted during simulations on earth. He took 37 minutes to perform tasks which previously had only taken a few seconds. An interesting feature of the flight was the accuracy of the splash-down – it was the first ever to be entirely plotted and initiated by the computer: they landed within two miles of the recovery ship.

The Gemini programme ended with Edwin Aldrin doing three spacewalks during Gemini 12, two of them lasting over two hours each. By improving training methods, cutting down the complexity of the EVA tasks, and providing more handholds on the spacecraft exterior, EVA was made less strenuous, and Aldrin was even able to do some gymnastics in space. Gemini 12 splashed down safely on 15 November, a fitting climax to the programme, and to an excellent year for NASA in 1966: as well as the 5 best Gemini missions, they had launched two Surveyors, two Lunar Orbiters and three Apollo tests.

Disaster The deaths of Komarov, Grissom, White and Chaffee remind the ambitious of the dangers of space flight, and cause second thoughts

After Gemini 7 Frank Borman spoke of the effect of publicity on the space programme. 'This programme is costing the American public a great deal of money and I feel it's our responsibility to report back to them . . . [but] when you expose a programme to the depths that our programme has been exposed you run several risks. For instance, any failure is magnified because several million people are watching it. . . . There's a tremendous amount of pressure to succeed. . . . I only hope that the people in this country are mature enough that when we do lose our first crew, they accept this as part of the business.'

For NASA 1967 was expected to bring the first manned Apollo flight, the first launch of the gigantic Saturn V rocket, and perhaps even the first manned lunar orbit. With enthusiasm deriving from the previous year's successes in the Gemini programme, there was even serious talk of a manned landing on the moon in 1968.

The Russians had launched no orbital missions since the March flight of Voskhod 2 in 1965. However, although there had been no manned Soviet flights in 1966, they had achieved both the first soft-landing on the moon, and the first lunar orbit in that year. Moreover the Russian calendar for 1967 held two highly significant dates: the 10th Anniversary of Sputnik 1 in October, and the 50th Anniversary of the Revolution in November. It did not take an expert in Kremlinology to guess that the big Soviet space come-back might be scheduled for one of those dates.

Today it seems that Soviet plans for 1967 were remarkably similar to those of NASA. They were preparing an unprecedentedly large (8 tons) one-man spaceship capable of docking in earth orbit with either an identical craft or a 'tanker' like the Agena. After technical trouble throughout 1966 they now planned to practise orbital docking during the first half of 1967, in the hope of being ready for a spectacular manned flight round the moon to coincide with the celebrations of the anniversary of the Revolution in November. Both Russia and America appeared in a position to achieve a lunar landing from mid-1968 onwards.

The name of the new Soviet spacecraft was Soyuz, the Russian word for 'Union'. To some it was taken as symbolising the progress of the Union of the Soviet Socialist Republics; to others it symbolised a mission where two sister spacecraft would join to proceed to the moon together.

The first important flight of the year was to be American – Apollo 1, in February; the second, Russian – Soyuz 1, in April.

Apollo 1 was to be the first manned Apollo mission. Between October 1961 and September 1964 there had been seven successful test flights of the Saturn 1 rocket, all part of the development programme leading to the Saturn 1 B, and eventually to its big brother, the Saturn V, the moon rocket itself. In February 1966 the Saturn 1B had flown for the first time, carrying a real Apollo capsule into a successful suborbital mission, which tested and proved its heat shield. The Saturn

1B was flown twice more that year, and the next flight, Apollo 1 was scheduled for 21 February 1967.

A month before that date the first Apollo astronauts White, Grissom and Chaffee were dead, the first victims of the space race. The supreme irony of their death was that they died within shouting distance of the ground, on top of an unfuelled rocket, going through an entirely routine test schedule. The public had been carefully prepared for death in space, for exploding rockets, eternal orbits, and faulty re-entries – but on 27 January 1967 no one was prepared for catastrophe. The rocket was surrounded by technicians on the gantry, and all the complex, emergency escape devices were completely useless. The escape rocket, the Malfunction Detection System, and the ingenious multiple duplication of controls had all been designed to cope with an emergency either in space, or during take off. As the flash fire raced through the Apollo capsule it was as though there were no safety precautions at all. Ironically, the test that day was due to end with a practice emergency escape from the capsule, anticipating a launch-fail situation. The escape had been timed so that the first man would be out in one and a half minutes. White, Grissom and Chaffee died in just 14 seconds.

At the time the spacecraft was full of pure oxygen at a pressure of 16 psi (normal atmosphere is 14·7 lb. per square inch), as it normally would be before launch. In orbit pressure was reduced to 5 psi. Oxygen is not in itself inflammable, but it is the gas on which all fire feeds: in chemical terms fire means oxidisation, and without an oxidiser fire cannot occur. In air not all materials will burn; in pure oxygen at 16 psi almost everything does. Plastic, perspex, any fibre, even metal, – all will burst into flame with amazing speed. The phrase 'to spread like wildfire' could have been invented to describe a flash fire in an oxygen-filled space.

In the Apollo disaster it was almost certainly a spark from badly insulated electrical wiring that started the flash fire. A less likely possibility was a spark from static electricity in the crew's clothing. They had been in the capsule since 1 p.m., and, after two hours of preliminary checks they signalled that they were ready to begin the 'dry run' countdown to the launch simulation, the object of the exercise. The hatches were closed, and at 2.50 p.m. the interior was pressurised with oxygen and the countdown began under real launch conditions. During the dummy countdown there were minor difficulties, and a bad crackle on the voice communication circuit, but the TV picture was clear. At 6.31, according to the official report, Chaffee was heard saying: 'Fire, I smell fire'. A second later there was movement inside the capsule, now assumed to be White initiating the emergency escape procedure. The TV screens flashed and went dead. One second after that instruments indicated a sharp rise in cabin temperature. After four seconds White reported: 'Fire in the spacecraft'. Then a third report, this time from Chaffee in a voice which left no doubt of the emergency: 'There's a bad fire in the spacecraft'. The cabin temperature soared. The radio went dead, and the oxygen supply to the astronauts' suits began to fluctuate wildly. The pressure rose, and 16 seconds after the fire broke out, the capsule burst, sending clouds of black smoke pouring into the upper levels of the gantry. The heat was so intense that engineers were not able to open the hatch from the outside for six minutes. Twenty-seven men suffered badly from smoke inhalation while trying to go to help the crew. Inside the charred bodies of the astronauts were found, one still strapped to his seat. Two hours later the Press was informed, and an all night search for the cause of the disaster began.

The whole of the colossal American Space Programme came to a halt overnight. Until it was certain that such a tragedy could never occur again, there was little chance of any resumption. For the first time since the end of the Mercury programme people began to ask fundamental questions about the objectives and methods of NASA.

The most obvious scapegoat was the NASA decision back in 1958 to use a pure oxygen atmosphere in the spacecraft. It had been a decision which recognised the risks, but those risks seemed to have been underestimated. What other similar deficiencies were there in this $15,000,000,000 programme? The day after the disaster a board of inquiry was appointed. Although it would have been inopportune to say it openly, many in NASA were conscious that November, the predicted time for a Soviet prestige flight, was only 10 brief months away; if the board of inquiry recommended abandoning the pure oxygen policy, it could be as long as five years before there would be any prospect of landing an American on the moon.

On 8 April the board of inquiry, the Thompson Committee, made its final report, a 3,000-page analysis of the Apollo disaster. As a semi-NASA report it was almost an orgy of self-criticism. Complacency was the root cause of the Apollo fire, and the report said so in no uncertain terms. 'Adequate safety precautions were neither established nor observed for this test', it said. As major direct causes of the accident the report listed: pressurised oxygen in a sealed cabin; extensive distribution of combustible materials in the spacecraft; easily damaged electrical wiring; combustible pipes carrying a glycol mixture; poor provision for the crew to escape; and inadequate arrangements for rescuing the crew and getting medical help.

The report inclined to the view that the spark originated in a bundle of wiring beneath a small equipment access door. Apollo contains 15 miles of wiring, and an inspection of it revealed 'numerous examples of poor installation, design, and workmanship'. The report referred to 'many deficiencies in design and engineering, manufacture and quality control'. One hundred and thirteen significant engineering orders were found not to have been carried out at the time the capsule was delivered to NASA. A wrench socket had been carelessly dropped among electrical wiring and left there; the astronauts had been allowed to sling nylon nets under their seats to hold miscellaneous articles – another additional fire risk. 'The Apollo team . . . in its devotion to the many difficult problems of space travel . . . failed to give adequate attention to certain mundane but equally vital questions of crew safety.'

The report recommended the elimination of all easily combustible materials from the spacecraft, the installation of a quick release emergency hatch, a general tightening-up of all fire precautions, and certain modifications to the wiring and the environmental control system. The major question of whether to continue with pure oxygen was side-stepped.

It was reported that several NASA executives suffered nervous breakdowns, but the task of modifying the capsule was undertaken with determination: never again would NASA be accused of complacency.

By September the new Apollo command module had been built, incorporating numerous fire precautions, and a quick release hatch which, assisted by a compressed nitrogen cylinder, could be opened in 5 seconds at the lightest touch.

The oxygen problem was partially overcome by a compromise solution: on the launch pad the cabin atmosphere would be 60% oxygen, 40% nitrogen, changing to pure oxygen during ascent. It was not until October 1968, with the success of Apollo 7 that NASA could consider itself back where it started – after a loss of 21 months. Long before that the Russians had their taste of tragic disaster.

After the Apollo catastrophe the Soviet trade union newspaper *Trud* said that White, Grissom and Chaffee were 'victims of the space race created by the American space programme chiefs'. It asserted that the American space programme had been characterised by 'haste': serious technical faults had been ignored, and the tragedy was 'far from being a pure accident'. Other Russian newspapers were more sympathetic with the Americans, and the world's Press remarked upon the open policy of NASA, and the complete lack of secrecy; unfortunately the full story of how the Russian tragedy happened has yet to be fully told. The basic facts that have been revealed are scanty.

The first manned mission using the Soyuz spacecraft was launched from Tyuratam in the USSR on 23 April 1967. According to Tass, the official Soviet News Agency, the 25-hour 18-orbit flight went entirely according to plan until the very last minute: Cosmonaut Komarov was fit, and fully in control during re-entry. Then, 'when the main cupola of the parachute opened at an altitude of 7 kilometres, the straps of the parachute, according to preliminary reports, got twisted, and the spacecraft descended at great speed which resulted in Komarov's death'.

There are many reasons for thinking that this report does not reveal the whole truth. It is extremely difficult to believe that a simple 18 orbits (perigee 125 miles, apogee 139 miles, inclination 51·7 degrees, period 88·6 minutes) was the intended mission of Soyuz 1. If so it would have marked no advance on the achievements of Voskhod 1 and 2 of two years before. It could have been simply a proving flight for the spacecraft, but it seems almost certain that Soyuz 1 was intended to rendezvous and dock with at least one other spacecraft. Speculation, conjecture, and rumour bedevil almost all Western accounts of Soviet space achievements, but, for lack of information, the writer has to make use of these admittedly highly suspect methods of ascertaining what really happened. In the case of Soyuz 1, there are unconfirmed reports that Valery Bykovsky was standing by for a second launch. It seems that there was no rendezvous, and that the Mission Directors at the Baikonur Cosmodrome finally decided to call the mission off, perhaps due to some trouble with either Soyuz 1 or the second launch vehicle.

The official report fails to explain why Komarov was unable to use his ejection seat and personal parachute, after he had realised that Soyuz 1 was in difficulty. Russian officials have neither admitted nor denied that Komarov even had an ejection seat. However, its absence would have been puzzling, for one recalls that even the multi-manned Voskhod capsule, whose landing system was a source of well deserved Soviet pride, probably had ejection seats fitted in case of emergency. Finally, however tragic, a parachute failure should have been a relatively simple matter to correct for future missions – within a matter of weeks one would have expected the fault to have been rectified, and the Soyuz programme to continue. However it was not until a year and a half later, in October 1968, that the Russians felt able to launch Soyuz 2 and 3. Whatever plans they had for celebrating the anniversary of the Revolution with a manned space flight achievement were dismissed. The inescapable conclusion is that something had gone drastically wrong. The Soviet manned flight programme was almost certainly held up by problems of the same magnitude as those restraining the Americans.

To assemble a reasonable explanation for the Soyuz tragedy, one is forced to sift conflicting opinion in an attempt to extract the most plausible account.

It would seem that the mission of Soyuz 1 was to provide a target for rendezvous, with Bykovsky in Soyuz 2, and possibly even a third ship of the same type. Soyuz 2 eventually was used as the name for a target vehicle for Soyuz 3, carrying Beregevoi. Having docked together, Soyuz 1, 2 and perhaps 3, would then have carried out complex manoeuvres and changes of orbit, similar to Gemini 10, 11 and 12 with their Agena targets. The flight would have lasted 7 days, ending triumphantly as part of Russia's annual May Day celebrations. Soyuz 1, with Komarov aboard, was launched at 3.35 a.m., Moscow time, and placed in roughly the planned orbit; preparations began at once to send Bykovsky to join Komarov 130 miles above the earth. However snags developed in Bykovsky's launch vehicle, and his countdown was held up. Before it could be resumed, Komarov reported gyroscope trouble. There were 3 gyroscopes on Soyuz 1, the first under the pilot's control, and the other two under ground control. The fault that developed could not be rectified by either Komarov or ground control. The gyroscopes always point in the same direction, no matter how the spacecraft moves in relation to the earth: they thereby provide a constant reference point for the spacecraft's attitude controls. Without them the spacecraft began to tumble wildly as it orbited the earth.

At lunch-time on 23 April Tass issued an unusually curt bulletin: it stated that the spacecraft had completed its fifth orbit by 10 a.m., and added that Komarov would be out of radio contact with Soviet territory from 1.30 p.m. until 9.30 p.m.

As dusk approached Mission Control had a difficult decision to make. Should they abandon now, and have the benefit of a daylight landing, or persevere, hoping for an improvement, and risk an emergency landing in darkness? They decided to continue. The News Agency was clamouring for information to release: normally Tass bulletins on flights are, if not very informative, at least regular. Shortly after midnight Tass was allowed to announce that Soyuz 1 had completed 13 orbits (actually it must have done this by 10.40 p.m.). The position was getting worse: maintenance instructions passed to Komarov had failed to cure the trouble, and the tumbling tendency increased. In attempts to steady the craft he made considerable use of the manual attitude control system, and fuel was running short. It became clear that there was little hope of carrying out the major objectives of the mission, and it was decided to bring Komarov down on his next orbit, the 16th. It was about 2 a.m. and by firing his retro-rockets he found he made the tumbling even worse. Re-entry under such conditions would have been suicide, since holding the craft with its heat shield forward would have been impossible.

The situation was very serious: the spacecraft would have to be steadied manually while Komarov attempted to initiate the complex re-entry procedure. Also, it was dark. By now he was in a lower orbit, and new parameters had to be calculated for his re-entry; it was too late to complete them for an attempt during the 17th orbit at 3.30 a.m. The next time round conditions seemed as good as they ever would be. Komarov fought to stabilise his spacecraft as it hit the upper atmosphere. An amateur tracking station in Turin claims to have picked up Komarov's side of the last conversation he had with Mission Control as he hurtled towards the Ionosphere. His words were: 'You are guiding me wrongly, you are guiding me wrongly . . . Stop! . . . Can you not understand? . . . Correct . . . Do you not remember. . . . Accelerate . . .'

The unfortunate cosmonaut was unsuccessful: Soyuz re-entered at the wrong angle, causing drastic overheating of the vulnerable parts of its equipment. In particular, the lid covering the parachute compartment became partially welded down as it was exposed to the searing heat of re-entry. When the radio command was given to set off the explosive charge which would blow the lid off, it stuck. A second explosive charge, an emergency one, was successful – but too late: Soyuz 1 was only 4½ miles from the ground, and still tumbling. The parachutes had no chance of deploying properly.

Astronaut Borman described his own re-entry in Gemini in the following terms: 'When you look at the nose of the spacecraft, it looks like an inferno, because the reaction jets are firing up there to keep the spacecraft stabilised; and then the phenolic material from the heat shield is boiling off and coming back over the windows, and impinging on the nose of the spacecraft, so that the front of you looks like a very holocaust – and when you realise that your parachute is up there you begin to wonder just how much heat it will take.'

Three minutes later Soyuz 1 hit the ground at 200 mph. If Komarov had not died from overheating, he was certainly killed by the impact, at about 5 a.m. An hour later Tass released the following bulletin: 'According to the report of the pilot cosmonaut Komarov, and telemetered information, he is in good health and feels well. The ship's systems are functioning normally.' It was not until 5.30 p.m. that radio and television simultaneously announced his death.

A few minutes after the announcement an obituary was published, signed by Brezhnev, Kosygin, Podgorny and other senior politicians and scientists. 'We shall for ever cherish the memory of the loyal son of our Motherland, wonderful Communist, courageous explorer of space, comrade in arms, and friend. He will be an example for ever of heroism, courage and valour. His name will summon the glory of our great Socialist country to new feats.'

The irony of Komarov's death was that, after becoming a cosmonaut in 1962, he had been dropped from the space programme because of an irregular heartbeat; it had taken considerable persuasion on his part to get the decision reversed.

Presumably there is a Russian equivalent to the Thompson report, giving full details of the flight of Soyuz 1. The file may also contain other names, although the rumours are in such mutual disagreement that it is difficult to take them seriously. The rumours mainly stem from the Russian magazine, *Ogoniok*, which, in October 1959, published a list of men supposed to be in training as cosmonauts. In it were the names of four men, Alexei Belokonev, Ivan Kaschur, Alexis Gratzev, and Jennady Michailov – they have never been heard of since. There have been various fanciful accounts of their deaths: it is alleged that Belokonev died on 12 November 1962 after a launch on the 8th. A Vostok launch is said to have failed in early October 1961, and a 2-man craft was rumoured to have been lost in space in June of the same year. A launch pad explosion in 1959 is said to have killed a cosmonaut called Andrei Mitkov. Even earlier, in late 1957, there is an unconfirmed report that all contact was lost with Alexis Ledovski when he was 200 miles up. Two other names often crop up in such stories: Serenti Shiborin and Piotr Dolgov. It is alleged in one source that Shiborin was killed in a space disaster in 1959, and elsewhere that he died in space in early 1958. Dolgov is rumoured to have piloted a spacecraft which was lost in October 1960: on that flight Dolgov was supposed to have intended to broadcast a message to the United Nations to coincide with Kruschev's visit to the USA. Yet another rumour states that Shiborin and Dolgov died together in company with an unnamed female cosmonaut on 22 May 1961. They were said to be on their 86th orbit, though at that date it is almost inconceivable that such a spacecraft could have existed. On 3 November 1962 a public announcement from Russia said that a parachutist called Dolgov had 'died on active duty'.

US Intelligence is said to have a list of 11 dead cosmonauts: 5 who failed to orbit, and 6 who died on the launch pad or during training. In an explosion in September 1960 Marshal Nedelin was killed, and many of the top executives of the Russian programme are said to have died with him. On 11 April 1961, the morning of Gagarin's flight, the British *Daily Worker* published a report from its Moscow correspondent, stating that a cosmonaut had died 3 days earlier during an unsuccessful 3-orbit flight.

It is quite possible that all these rumours are 'wicked imperialist lies'; however, if they are, they are not included in this book as malicious slander, but merely as attempts to fill gaps in the history of manned space flight which could be filled much more easily if the Russians would open their files.

The dangers of space travel are constantly being emphasised, yet, compared with conventional aviation, space travel has a good record so far; at least one cosmonaut and six astronauts have died in air crashes, whereas, only four are known to have died in spacecraft, and there is no official record of anyone being killed in outer space.

The danger, nevertheless, is very real: Michael Steiko, an advanced-concepts engineer with Martin Co., has calculated an estimate based on statistical probability, assuming that both Russia and America continue their space programmes along established lines for the next twenty years. In that time he reckons there will be about 380 manned space flights, involving 400 American astronauts and 400 Russian cosmonauts. He calculates a 62% probability of at least seven emergency situations involving 22 men.

The test flying of a conventional aircraft can be carried out over months, with each flight becoming progressively more ambitious – and fuel being the only major expense. However each time a spacecraft is launched, it costs several million dollars: because of this every flight is designed to incorporate the maximum advances over the previous one, and the risk element grows proportionally.

Decompression is the first danger, caused by a failure in the oxygen supply, or by a puncture of the capsule by a meteorite. Spacesuits provide a secondary protection in the event of cabin pressure loss. In Apollo one astronaut is nearly always in his suit, in order to save the others by either repressurising the cabin, or helping them into their suits.

Explosive decompression can occur when the suit itself becomes punctured – in EVA for example. If the loss of pressure is total, and happens in less than half a second, death is certain due to the vaporisation of all the liquids in the body. If the decompression is slower, and the body brought back to a pressurised atmosphere within 80 seconds there is a good chance of survival. Vaporisation begins in the heart and lungs within 6 seconds of losing pressure; the lungs then collapse and the blood pressure drops, causing acute anoxia, convulsions, and distention of the bowels. The heart stops beating after 2 minutes. If the astronauts were breathing air as opposed to pure oxygen the depressurisation would inevitably be fatal, for they would suffer from acute dysbarism (or the 'bends' as it is known to deep sea divers).

For many years it had been known that outer space almost certainly held an invisible threat to man in the form of cosmic radiation – streams of high energy particles which originate from our own sun and distant galaxies, and travel at the speed of light. Life on earth is sheltered from this cosmic hailstorm by our atmosphere, and earth's magnetic field: they absorb and deflect this cosmic radiation. However, before man could venture into outer space, satellites had to be sent up to investigate these radiation hazards. That was one purpose of the early Sputniks, the Discoverers, Solrads and Explorers. They carried particle counters and relayed their findings back to earth. Cosmic radiation was found to be lower than expected, and with a certain amount of shielding man could be adequately protected. However, it was also confirmed that radiation increased to dangerous levels during a solar flare. High energy flares occur every four or five years and last several days. Medium or low energy flares happen several times every year.

Fortunately it was found that by keeping a close watch on the sun, one could predict the more devastating surges of radiation, giving the spacecraft time to turn its thickest shielding towards the sun until the danger was over. 'Solar Alerts' will be a feature of space travel for many years to come. Also it was found that, quite apart from the normal surges of radiation, with which a spacecraft could be designed to cope, there were occasional block-busters made up of heavy particle concentrations. In these the particles are a hundred times more energetic than normal, and against such heavyweights there still remains no defence, for adequate shielding would make any spacecraft too heavy to launch.

Statistically the chances of a spacecraft hitting one of these radiation block-busters are very small, but they may well remain one of the major hazards of long duration space flight.

The early satellites made another, but quite unexpected discovery which came as a big shock to science. When the 31-lb. satellite Explorer 1 was launched into a high elliptic orbit on 31 January 1958 it sent back readings during parts of its orbit which could not be explained as any hitherto known radiation source. A picture gradually built up of local areas of high radiation quite close to the earth. Explorer 1 stopped transmitting on 23 May, but by then sufficient information had been received to understand the remarkable phenomenon it had discovered. Over the centuries countless millions of electric particles had become 'trapped' by the earth's magnetic field. Oscillating between the poles they formed 2 belts of intense radiation around the earth. They were named the inner and outer Van Allen belts. Both the Russians and the Americans carried out research programmes to map these belts, and nearly every space vehicle launched in 1959 and 1960 carried a radiation experiment. It was found that, passing through both belts with some degree of screening protection, an astronaut would receive a radiation dose of about 5 rem. A fatal dose would be about 500 rem. The belts do not therefore present an insuperable barrier, but they could make problems for the positioning of permanent orbiting space stations.

Another danger is radiation from high altitude nuclear tests. The Chinese exploded a high altitude device on the very day that the Apollo 8 capsule was due to re-enter the atmosphere and pass over China. Borman, Lovell and Anders seemingly suffered no ill effects, but it could have put them in extreme danger.

The most publicly familiar danger to astronauts and cosmonauts alike is a collision with a meteorite. Our atmosphere protects us from them, but they range in size from tiny dust particles to chunks of rock weighing several tons. At one time it was thought they would make space travel impossible, but a degree of protection is feasible. The dust particles, or micrometeorites are extremely common, but so small that they do no damage. Large meteorites from an ounce upwards, represent a calculated risk which can be ignored, although an unlucky collision would mean certain disaster.

The chances of an encounter with a medium-sized meteorite, weighing up to one-tenth of an ounce, are appreciable. So far it has been found possible to protect spaceships against them by building double walls: the meteorite impact releases sufficient energy to melt part of the outer wall, making a shock crater on the inner surface. Thus the inner wall stays intact, and today advanced spaceships have a self-sealing compound at pressure between the walls to repair any damage automatically. Astronauts even carry 'puncture kits' to repair the tiny holes made by unusually fast meteorites, but so far no American has had to use one. In 1965 some of the Apollo test flights were used to launch three huge Pegasus satellites as part of a meteorite study programme. These Pegasus satellites deployed vast wings, 2,900 sq. ft. in area, and over one year recorded 1,100 impacts between them; the meteorite problem is far from solved.

On 22 April 1968 a United Nations pact was signed by the USA, USSR, UK, and 19 other countries. It bound them to render 'all possible assistance to astronauts in the event of accident, distress, emergency, or unintended landing'. But true international cooperation in the conquest of space still seems a long way away: Frank Borman, in a BBC interview on this topic, said: 'Well, this is a very idealistic thing – we would all like to cooperate in many fields. Unfortunately we haven't found the Russians too cooperative in any field, and I'm afraid that the language barriers, and just the mechanical difficulties of cooperating to . . . any great extent, would be very, very difficult. We have difficulty sometimes in getting North American, who build one part of our spacecraft, to cooperate with Grumman, who build the other part. I can't imagine what it would be like if we had communications between Moscow and Washington.'

As Borman inferred, the technical sophistication of American and Russian spacecraft would make any rescue operation by the other side a virtual impossibility. A few weeks before his death, Ed White made this poignant comment on the dangers of space exploration: 'As we fly more and more spacecraft, we are . . . probably going to lose somebody, but I wouldn't hold up the space programme on that account.'

ⅰbove *Moscow, 26 April 1967. ʼhe urn containing the ashes ƒ Cosmonaut Komarov, eing carried by Gagarin, ʼopovich, Nikolayeva-ʼereshkova, and Titov.*

Apollo The project to land an American on the moon

'The greatest scientific, engineering and exploratory challenge in the history of mankind' – that is how the United States Apollo programme has been described, and this year, 1969, saw its climax: the landing of Neil Armstrong and Edwin Aldrin on the moon and their safe return to earth. Perhaps the most extraordinary element of the story is the speed with which America has achieved its objective. 'No single space project in this period will be more exciting, or more impressive to mankind.' That is how President Kennedy summed up the American lunar landing mission in his 'in this decade' speech to Congress only eight years before when he stated the national objective of taking 'a clearly leading role in space achievement'. He left some Congressmen incredulous by prophesying that America would get a man on the moon by the end of the decade. For the self-avowed 'greatest nation in the world', a nation suffering considerable 'loss of face' after the achievements of Sputnik and Gagarin, Kennedy's speech was, as we have seen, a shot in the arm. Admittedly twenty days before the speech Alan Shepard had boosted American morale by becoming the first American to get into space, but his feat hardly compared with Gagarin's earth orbit in the previous month. Overnight it had been assumed that the Russians wanted their man on the moon first: the few critics who suggested that unmanned expeditions would be cheaper and equally worthwhile and that the 'long range exploration of space' could be advanced further in other ways, went unheeded. American prestige was at stake, and the race to the moon was on.

The actual Apollo programme itself had been announced by NASA in July 1960, but it was the Kennedy speech that put it into top gear, and by the end of November 1961 the basic Apollo spacecraft contract was awarded to the Space Division of North American Aviation Inc. (which merged with the Rockwell Standard Corporation in 1967 to become North American Rockwell).

At the time of Kennedy's speech most objective observers regarded America as years behind Russia in the space race; yet now she was proposing not only to put men up in earth orbit, whence they could presumably return within minutes in an emergency, but to put them farther into space for whole days at a time: it could be as long as five days before men going to the moon could get back to earth. This meant a capsule in which three men could not only survive for two weeks, but carry out control duties, a capsule which could be absolutely guaranteed to go on working whatever happened, and return them safely. The other problem was the launch vehicle: could a propulsion unit be built which would be sufficiently safe, yet powerful enough to blast three men and their complex retrievable 'environment' to the moon?

As we have seen, several methods of reaching the moon had been considered; the favourite was the direct ascent trajectory, whereby everything needed for the trip to the moon and the return is put into space at one go on a single monster launch vehicle. It was not until July 1962 that NASA revealed its final plans. They chose the lunar rendezvous technique. This meant putting a two-man vehicle on the moon. This 'lunar module' would ferry the men back to a spacecraft parked in orbit around the moon. The lunar module would then be jettisoned before the spacecraft made the return journey to earth.

This decision meant that three men would have to be put into orbit around the moon, the third remaining in orbit while the other two made the final descent to the moon surface to spend up to a day and a half there before rejoining the orbiting spacecraft. From launch to splashdown the whole trip would last between 8 to 10 days, but the spacecraft had to be designed for at least a 14-day operation to give the crew a margin of safety. To make that fortnight in space possible has involved the wealthiest nation in the history of mankind in the biggest research programme ever undertaken.

In 1957 Wernher von Braun was carrying out studies aimed at developing a rocket booster with a total thrust of 1,500,000 lb.: many still regarded his ideas as science fiction, yet four years later the rocket was successfully launched. Late in 1958 the development of such rockets was still a military monopoly, and work began at the Redstone Arsenal in Huntsville, Alabama, conducted by the Army Ballistic Missile Agency. In 1959 the name Saturn was given to the 1,500,000-lb. launch vehicle. Then came Saturn 1B, a two-stage launch vehicle: it was first launched early in 1966 carrying the Apollo spacecraft for its initial test in space. The last member of the Saturn family is the moon rocket itself, the Saturn V. It is America's largest and most powerful launch vehicle. Fully loaded, the Apollo Saturn V lunar vehicle stands 363 feet tall – that is, 2 feet less than St Paul's – and weighs more than 6,000,000 lb. The engines of Saturn V launch vehicle which propel the Apollo spacecraft to the moon have a combined horse-power equal to 543 jet fighters at full throttle.

The first stage is 138 feet high and weighs nearly 5,000,000 lb. Its function is to lift the second two stages and the spacecraft itself to 41 miles above the earth and to accelerate them to 5,400 mph. At that moment, about two and a half minutes after leaving the launch pad, the five F-1 engines cut off, and retro-rockets on the first stage start to fire to decelerate the first stage and allow the second stage to separate and continue its journey. Four seconds later the five J-2 engines of the second stage take over, boosting the spacecraft and the third stage in about 6 minutes to a speed of more than 14,000 mph and an altitude of about 120 miles. This second stage is 81 feet tall and weighs more than 1,000,000 lb. when fuelled. It is the largest, most powerful hydrogen fuelled stage ever made. Thirty seconds after the ignition of the second stage, at a height of about 295,000 feet the astronauts jettison the launch escape structure. This safety device perches on the very top of the whole Saturn V complex. It is 33 feet long, and attached by a lattice work tower to the command module, which contains the astronauts.

In the event of an emergency on the pad or shortly after launch, the launch escape subsystem carries the command module sufficiently high and to the side of the main launch vehicle for the command module's own parachutes to bring the crew safely back to the ground. For maximum dependability at all times its three rocket motors use solid propellant; the command module is shielded from its exhaust by a protective cover attached to the base of the tower. This cover also protects the command module from the heat generated quite normally as the launch vehicle accelerates up through the atmosphere; it is jettisoned with the rest of the launch escape assembly.

An emergency detection system automatically activates the escape assembly during the first 100 seconds, and the astronauts can also operate it manually at any time from waiting on the launch pad right until the moment they finally decide to jettison it.

At an altitude, then, of 120 miles, the third stage takes over after the second stage has been jettisoned. It accelerates the spacecraft to over 17,000 mph, putting Apollo into an earth orbit. During the ascent the crew have been watching their controls ready to abort the mission if all systems are not functioning perfectly. In orbit they have a chance to check all subsystems before committing themselves to the final journey from the earth to the moon, the translunar insertion. They can orbit the earth up to three times before inserting themselves into translunar flight, but because this insertion is desirable as soon as possible after confirming that 'all systems are go', they normally intend to begin this manoeuvre on the second orbit.

The moon may appear an easy target to the earth-bound observer as he gazes at its enigmatic face in the night sky. To the astronaut, however, a moon shot demands the highest precision and accurate timing. He must approach it on a figure-of-eight type of trajectory so that, if for any reason the spacecraft does not achieve an orbit around the moon it will at least loop it and eventually make a 'free return' to earth. If such a trajectory is not achieved the spacecraft would continue past the moon, into an eternal orbit of the sun. To help the astronauts with this decision the Saturn V third stage is equipped with an instrumentation unit which computes the exact moment when they should make their bid for the moon: the crew then re-ignite the single J-2 engine of the third stage. It burns for about five and a half minutes, cutting off at an altitude of about 190 miles, at a speed of 24,500 mph.

Back on earth the vast network of the manned space flight mission tracks the spacecraft for about 10 minutes after the third stage has cut off in order to confirm the correct precision of the flight path. The astronauts are then told to begin the complex docking manoeuvres which prepare the spacecraft for its arrival and landing on the moon.

With the launch escape assembly jettisoned before the ignition of the third stage, the command module containing the three astronauts now tops the vehicle. The command module (CM) acts as both control centre and living quarters for most of the lunar mission. Its pilot remains in it for the entire mission and the other two only leave it for the excursion to the moon surface. It is a roughly conical structure, 10 feet 7 inches high and 12 feet 10 inches in diameter.

Below it sits the service module (SM) containing critical subsystems and fuel supplies for almost the entire lunar mission. It remains attached to the command module from launch until just before re-entry. The lunar excursion module itself, the only part of the whole complex which will actually land on the moon, is contained inside a protective aluminium structure called the Spacecraft-Lunar Module Adapter: this protective structure sits between the spacecraft and the third stage of the launch vehicle.

It is at this moment, drifting towards the moon, that perhaps the most remarkable operation in the whole extraordinary journey takes place. The crew position themselves to see clearly out of the docking windows. Then the commander fires the engines of the service module; almost at the same time the panels of the spacecraft-lunar module adapter are jettisoned: the command service module is stopped after it has travelled about 60 feet away from the third stage. The commander then turns the CSM through 180 degrees until he can see the lunar excursion module which is still attached to the third stage: he now lines up the command module with the lunar module and returns, docking the top of his command module into the roof of the lunar module. After checking that all the docking latches are engaged, the spacecraft with the lunar module attached is then separated from the third stage by spring thrusters, and the 240,000-mile journey to the moon continues. At over 24,000 mph one would think it would only take 10 hours to get there, but in fact it takes about 66. This is due to the influence of earth's gravitational field which continues to slow the spacecraft to a point 30,000 miles from the moon, when it enters the moon's sphere of influence at a speed of 2,120 mph. The precise location of the exact balance between the gravitational fields of earth and moon is known as the Neutral Point (NP). Past this point the spacecraft accelerates under the pull of lunar gravity.

During transit the spacecraft is navigated by the stars, and also, perhaps surprisingly, by using landmarks on earth. To protect the craft against the extremes of temperature in outer space, it is put into a slow roll, revolving twice an hour to provide uniform solar heating.

Beyond the NP the spacecraft accelerates towards the moon, reaching 5,600 mph; it is slowed down by a retro-firing of the main engine to 3,600 mph to go into an elliptical orbit ranging from 70 miles to 196 above the moon surface. Approximately two revolutions later this manoeuvre is repeated to convert the orbit into a circular one to keep the spacecraft at a regular 70 miles above the surface. This braking manoeuvre is carried out by the Service Propulsion System (SPS), and because precise timing is again vital, the exact moment of ignition and the duration of the firing are determined by the Mission Control centre in Houston and then programmed into the command module computer which automatically fires the engine.

Once in lunar orbit the crew compare tracking data from earth with landmarks they can actually see below them on the moon surface. They can then work out exactly where they are and make changes in heading, to place them above the desired landing site. (Apollo 8 was the first mission to show that American astronauts had become experts in moon geography.) The lunar excursion module and the tunnel to it from the command module are now pressurised, and the two hatches removed. The lunar module pilot can now move from his couch in the command module and float through the narrow tunnel to enter the lunar module: he has practised this many times on earth, but in space it is much easier because of his state of weightlessness. Having switched on the module's environmental control, its electrical power, and communications subsystems, he is followed by the Mission Commander; together they check the lunar module subsystems and prepare for separation. Meanwhile back in the command module the CM pilot is making more sightings and completing the alignment and attitude adjustment of the spacecraft.

The two hatches are now closed, the 12 docking latches released, and the lunar module's own reaction control system fired to separate the two modules. The LM is positioned to enable the astronaut who stays in orbit in the CM to inspect the LM landing gear which was deployed just before separation. If he gives approval the LM can begin its descent: during this final stage of the journey to the moon both the CM and the LM have to move round behind the moon.

Communications with earth necessitate line-of-sight, so for the 45 minutes spent behind the moon out of each 2-hour lunar orbit, the astronauts are on their own, completely out of touch with Mission Control. During that period equipment aboard the CSM records all conversation and data exchanged between the LM and the CSM, and relays it back to earth when they emerge from behind the moon.

The actual descent takes about an hour: first the LM has to manoeuvre itself into an elliptical orbit that reaches from 70 miles to within 50,000 feet of the moon's surface. At the closest point to the moon on this orbit the LM's descent engine is fired again to slow down the module. At 9,000 feet the landing site should come into view, but the firing is controlled automatically until the craft is only 500 feet above the moon surface. At that point the commander takes over and tilts the craft in order to get a good view of the landing site.

If the commander has second thoughts about landing, this is his last chance, for at 65 feet he must re-orient the LM in order for it to descend vertically to the surface at about 3 feet a second. As soon as the landing gear touches the moon he shuts off the descent engine. Only half of the LM returns to the CSM orbiting 70 miles above. Before exploring their landing area the commander and LM pilot spend about 2 hours checking out the ascent stage of the module to guarantee their safe return. Another 2 hours are needed for checking their lunar suits, and extravehicular mobility units.

Finally the LM cabin is depressurised and one of the crew emerges from the module to descend its ladder.

The LM pilot records the event for posterity using both cine and still cameras; he films from the ascent stage of the module, recording the activity of the first 20 minutes as the Mission Commander gathers samples of surface material for later analysis. Then, having unloaded research equipment, the two astronauts inspect the LM to see the effects of their lunar touchdown. If all is well they then make a TV scan of their landing site and begin a collection of rock samples. After a sleep period they continue their exploration of the moon, but on early missions will probably only venture up to about a quarter of a mile away from the LM. They have plenty to do, for quite apart from photography and sample gathering, they are scheduled to build a station that will continue sending scientific data back to earth after their departure.

To leave the moon they climb back into the ascent stage of the LM. The descent stage remains on the moon, acting as a launching platform. The engine of the ascent stage lifts them to 60,000 feet from the surface where they go into an elliptical orbit reaching 34·5 miles from the moon at its farthest point.

The next manoeuvre problem facing all three astronauts is formation flying at different altitudes. The LM has to change from an elliptical orbit to a circular one, positioning itself at a constant altitude below the CSM. Then the LM fires its reaction control engines to lift it up to the 70-mile orbit of the CSM. During this part of the return journey the CM pilot has the task of tracking the LM below; the interception takes about 30 minutes, and the course corrections are controlled by a computer in the LM on the basis of data fed from earth. When the commander considers the LM sufficiently close to the CSM he takes over control and manoeuvres it with short bursts of the reaction control engines until it docks back on to the top of the CSM. Pressure in the two capsules is equalised, and the hatches removed to give access to the tunnel joining them. After the lunar samples, film, and equipment needed back on earth have been transferred to the CM, everything that is surplus to the return journey is dumped in the LM, the hatches closed and the seal checked. Small charges around the CM docking ring are fired, separating the entire docking mechanism from the CM, and jettisoning the LM. A short burst with the reaction control engine of the service module moves the CSM away from the LM, leaving it in a lonely orbit around the moon.

The CSM now has to accelerate to escape the moon's gravitational pull and regain the earth's sphere of influence, where it is able to coast home at ever-increasing speed. At this moment of acceleration it will again be behind the moon, out of communication with earth. The engine is fired for about two and a half minutes, boosting speed from 3,600 to nearly 5,500 mph. The re-ignition of the service propulsion system is one of the main danger points of the lunar flight; it means anxious minutes on earth before the craft emerges from behind the moon to reveal whether or not the astronauts are on their way home. At the moment of reappearance on Apollo 8, Jim Lovell echoed everyone's thoughts when he said: 'Please be informed there is a Santa Claus.'

The return trip is the longest part of the mission, lasting up to over 100 hours; when the spacecraft re-enters our atmosphere its speed is in the region of 25,000 mph. Such velocity generates a temperature of about 5,000 deg. F. on the skin of the capsule, but the cabin should remain at 80 deg. The service module is jettisoned just before entry into the earth's atmosphere. During separation the service module's reaction control engines are fired to ensure that the two modules do not collide. Also, about three and a half hours before entry the whole CSM is rotated to put the forward heat shield of the CM in the shade, cooling it for an hour and a half before the final course corrections.

The Apollo heat shield is made of a special stainless steel honeycomb sandwich with a phenolic-filled epoxy compound as the heat-dissipating, ablative material. At its thickest point, where the forward edge of the base of the CM first hits the atmosphere it is 2·75 inches thick. From this point the thickness tapers off, so that the aft part of the base is only 1·5 inches thick, and the top of the capsule a mere ·75 inches thick. For additional heat protection this outer structure is joined to the inner pressure shell by a layer of fibrous insulation.

Re-entry, when the CM begins to feel the resistance of the earth's atmosphere, begins at about 400,000 feet. Soon after this, the ablative coating turns white hot at 5,000 deg. F., charring and melting away, carrying most of the heat with it. The whole landing is controlled automatically by an earth landing subsystem, although the crew have backup control. At 24,000 feet a barometric switch starts the subsystem, the forward heat shield is jettisoned, and drogue parachutes are released; they slow the module from 325 to 125 mph, and steady it in readiness for the opening of the main parachutes.

Just above 10,000 feet these drogues are disconnected, and pilot parachutes pull out the main chutes to slow the CM to 22 mph ready for splashdown.

The module is suspended from the parachutes at an angle which enables the front edge of the base to hit the water first. Four crushable ribs of corrugated aluminium crumple on impact with the water, and greatly reduce the shock of the landing. The crew are doubly protected from a severe blow by the struts supporting their couches: inside the struts, pistons absorb the energy by deforming steel wire rings in a manner similar to the steering column of some cars, which telescopes during a bad crash.

Just before the Apollo 11 team set off on their epic journey, President Nixon called their excursion into the unknown 'man's greatest adventure'. In this Chapter, I have described the technical plan behind that adventure. In 1968 the plan was still only an audacious theory. In July 1969 it became history. It was a project so brilliantly conceived that, although it contained many 'life critical points', it gave the astronauts at least as good a chance of returning as Sir Edmund Hillary had when he conquered Everest.

Right *Dr Wernher von Braun following the countdown of an early Apollo mission.*

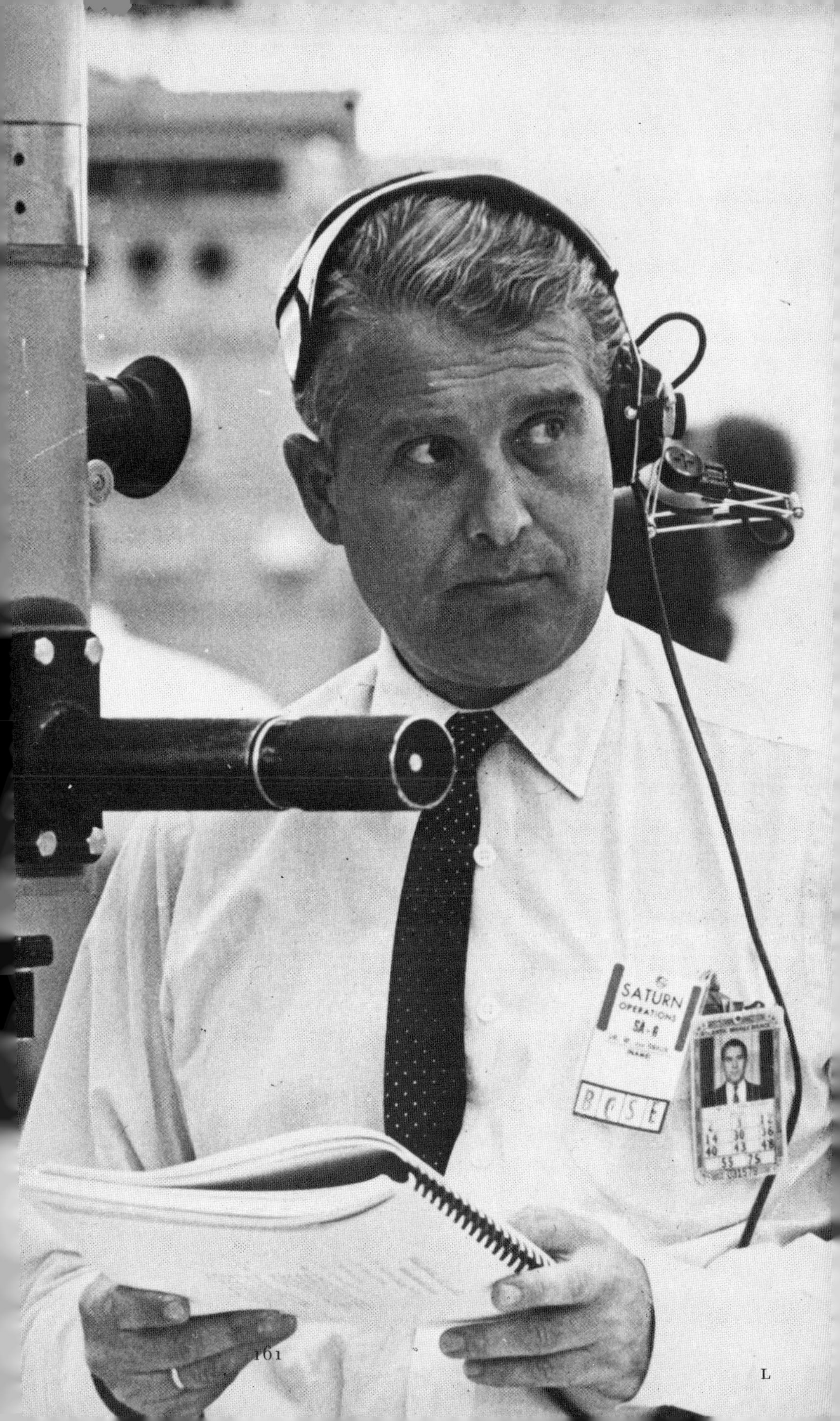
SATURN
OPERATIONS
SA-6
BOSE

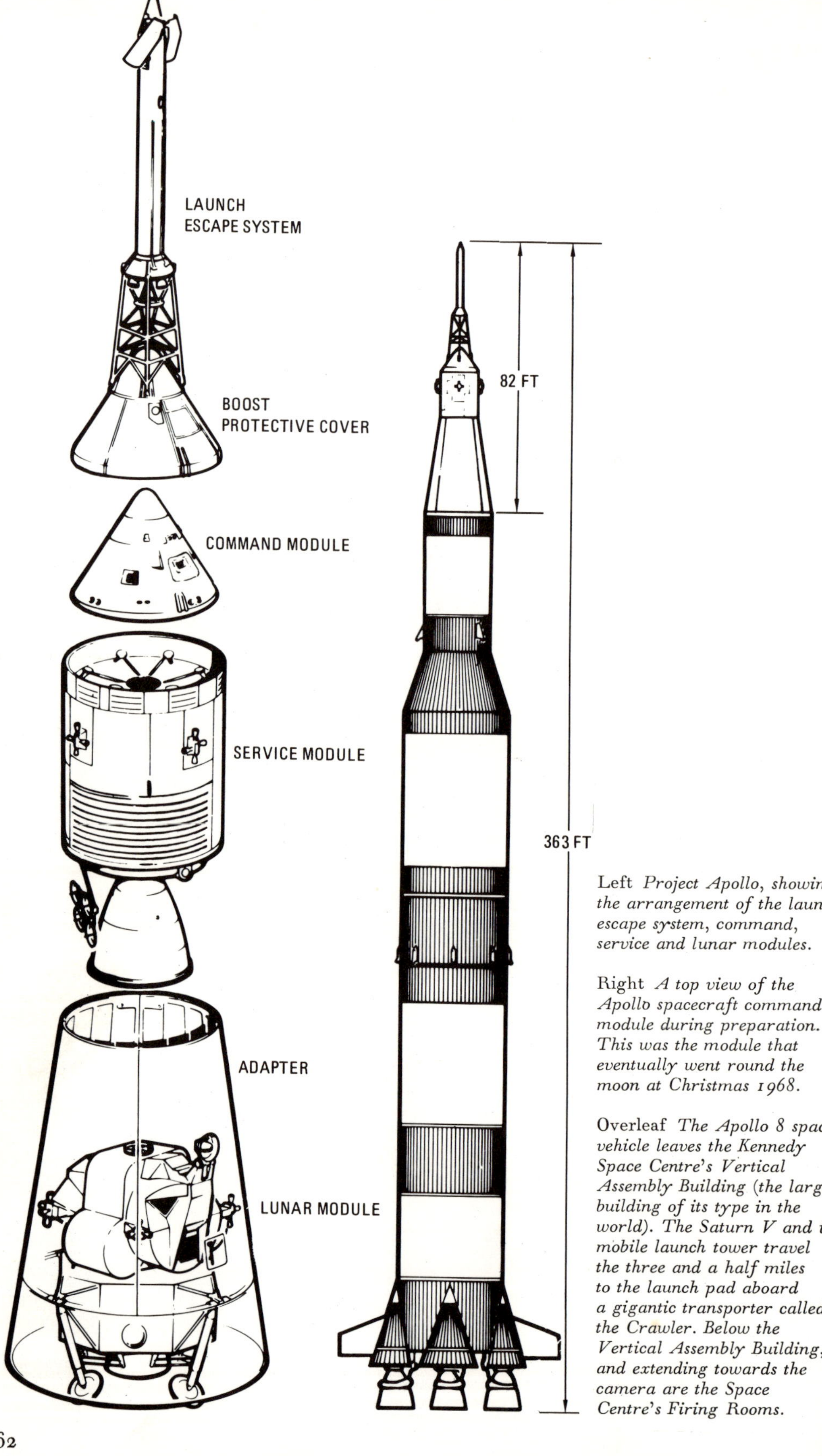

Left *Project Apollo, showing the arrangement of the launch escape system, command, service and lunar modules.*

Right *A top view of the Apollo spacecraft command module during preparation. This was the module that eventually went round the moon at Christmas 1968.*

Overleaf *The Apollo 8 space vehicle leaves the Kennedy Space Centre's Vertical Assembly Building (the largest building of its type in the world). The Saturn V and its mobile launch tower travel the three and a half miles to the launch pad aboard a gigantic transporter called the Crawler. Below the Vertical Assembly Building, and extending towards the camera are the Space Centre's Firing Rooms.*

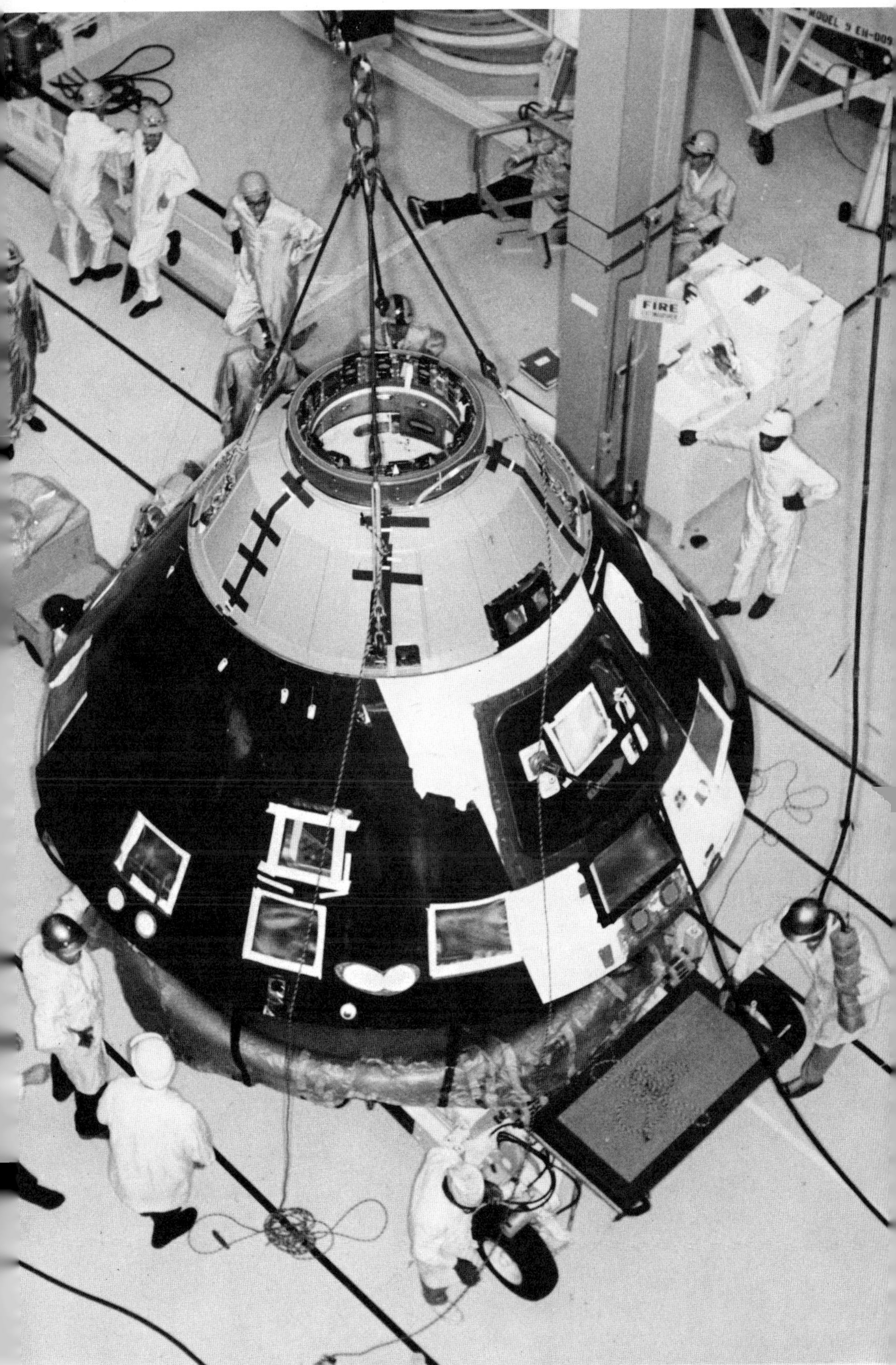
FIRE

USA

Above *One of the Kennedy Space Centre Firing Rooms, here seen during the Apollo 4 mission. Each room has 450 consoles containing controls and displays required for the checkout process, plus fifteen display systems, each capable of providing digital information instantaneously.*

Right *A view from the Vertical Assembly Building of the giant Saturn V rocket with its mobile launch tower aboard the Crawler. 65 television cameras are positioned around the Apollo vehicle to monitor every launch.*

USA

USA

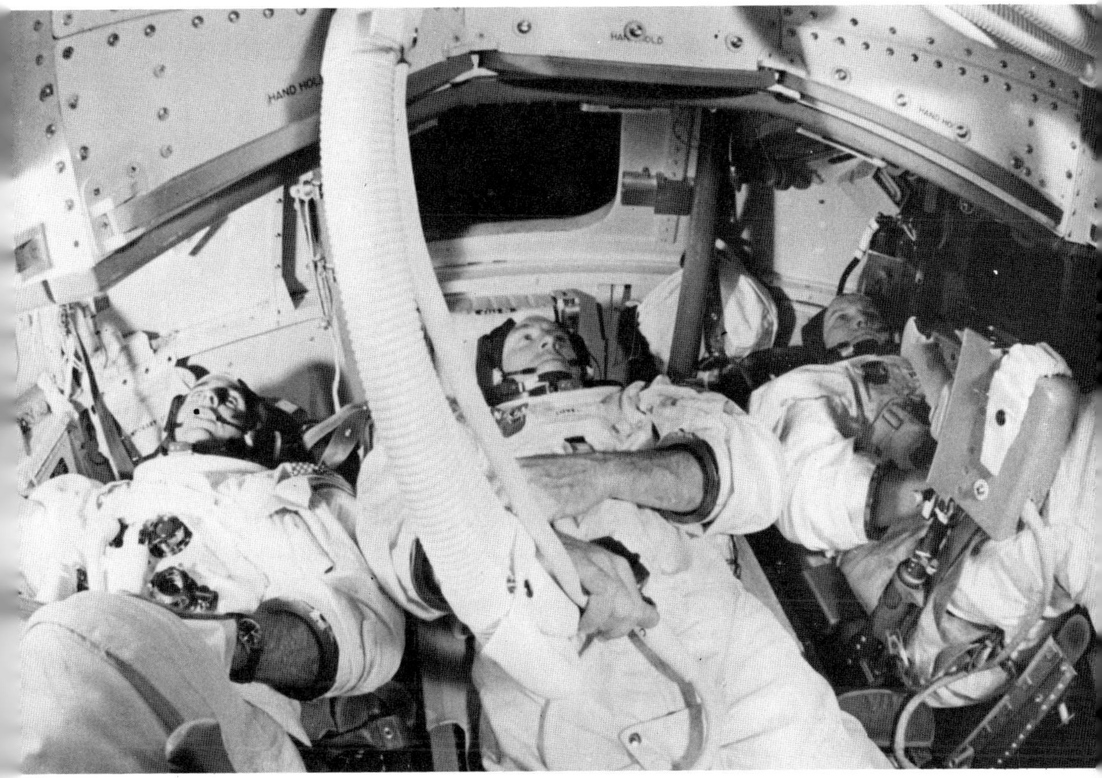

Left *The launch of Saturn V.*

Above *The Apollo 8 crew inside a simulator, training for their lunar orbit mission. Left to right: William A. Anders, James A. Lovell Jr. and Frank Borman, commander.*

Right *Cross-section of the Apollo command module showing the various work and sleep positions.*

Overleaf *An artist's impression of the separation of the Saturn first stage on the flight to the moon.*

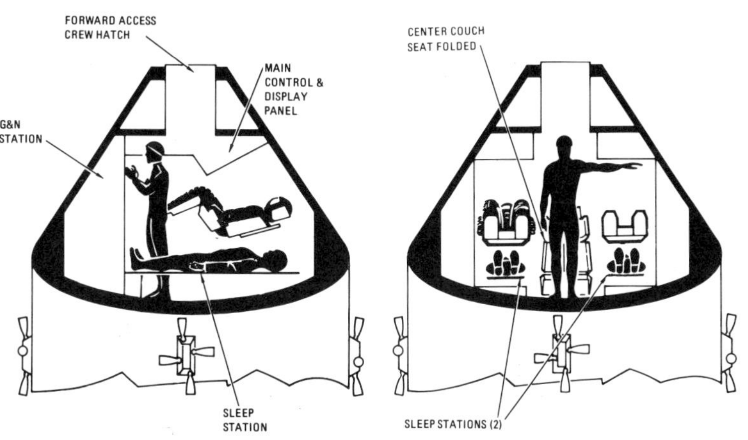

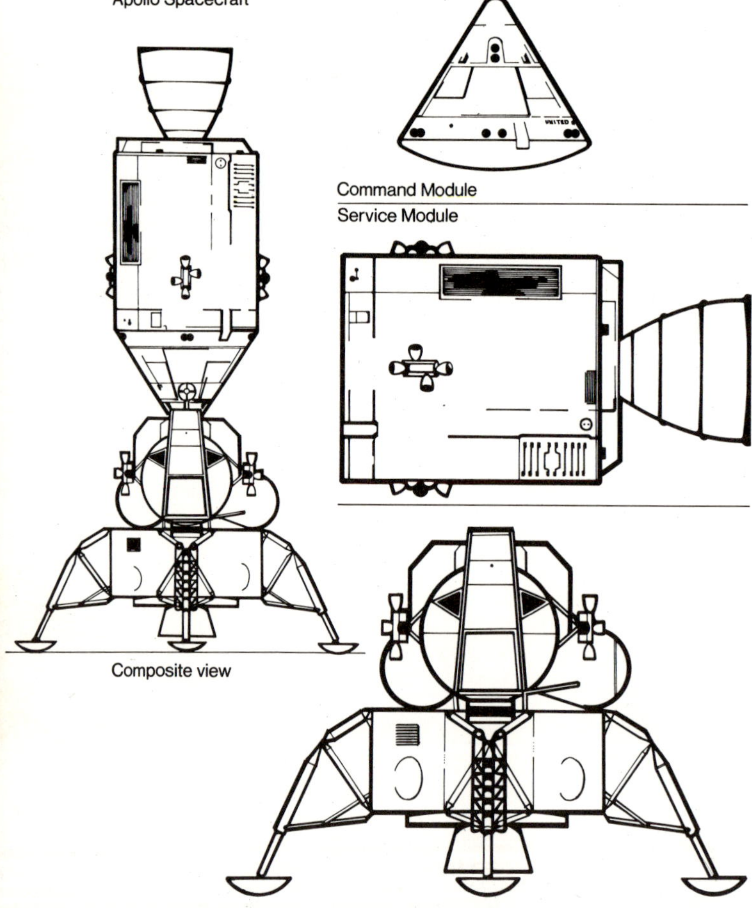

Above *An artist's impression of the second stage separation.*

Left *The basic components of the Apollo spacecraft.*

Above right *The command service module leaves the Saturn V third stage to begin its docking operations with the lunar module.*

Below right *Apollo 8 view of S-IVB. A photograph taken from the Apollo 8 spacecraft looking back at the Saturn V third (S-IVB) stage, from which the spacecraft had just separated following translunar injection. Attached to the S-IVB is the lunar module test article (LTA) which simulated the mass of a lunar module on the Apollo 8 mission. The 29-foot panels of the spacecraft LM Adapter which enclosed the LTA during launch have already been jettisoned and are out of view.*

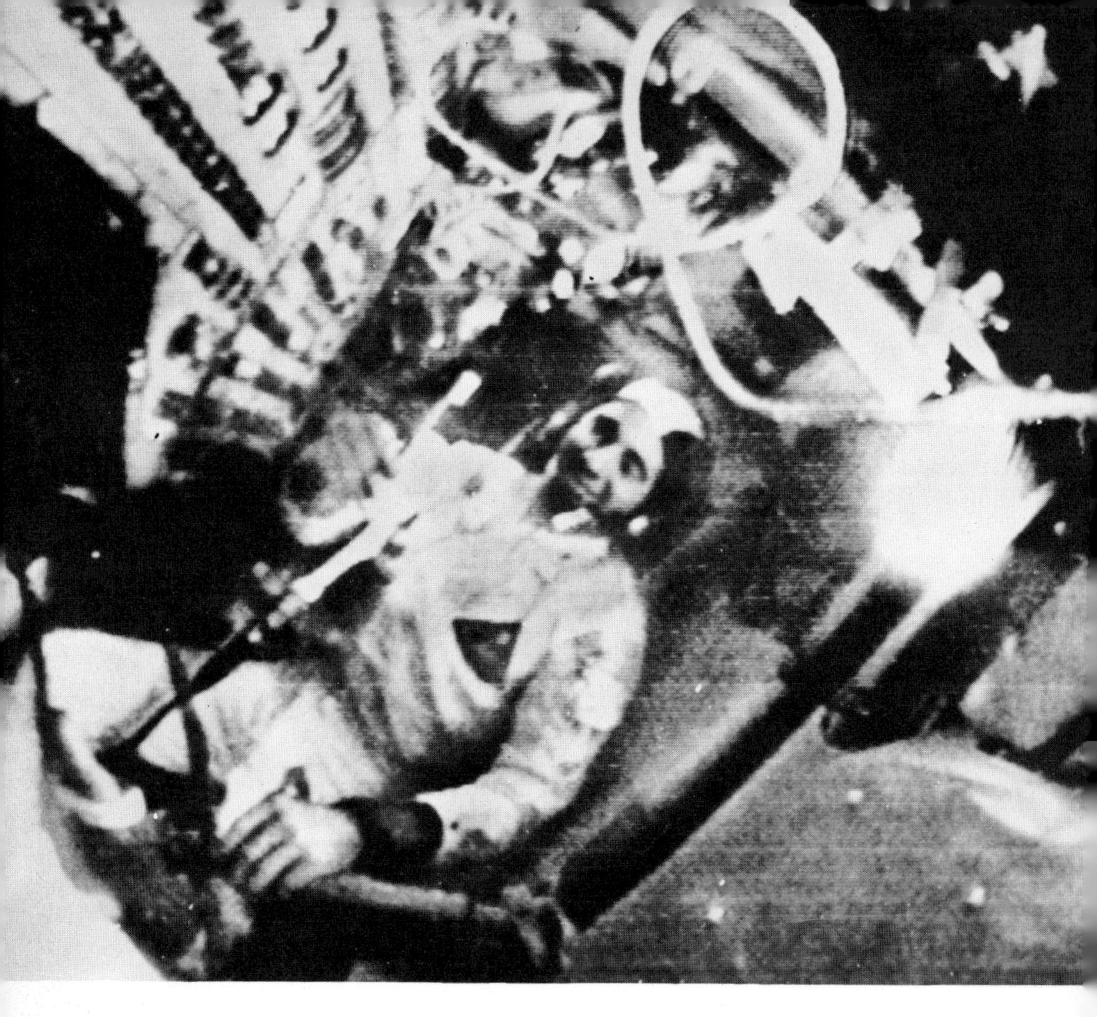

Above *Apollo 8 astronaut William Anders demonstrates weightlessness with his toothbrush to television viewers during the first live transmission from the spacecraft on 22 December 1968, when he was already half way to the moon, 139,000 miles from earth.*

Right *The CSM is turned to dock with the lunar module.*

Overleaf *Cape Kennedy as seen from Apollo 7, with the S-IVB in the foreground.*

M

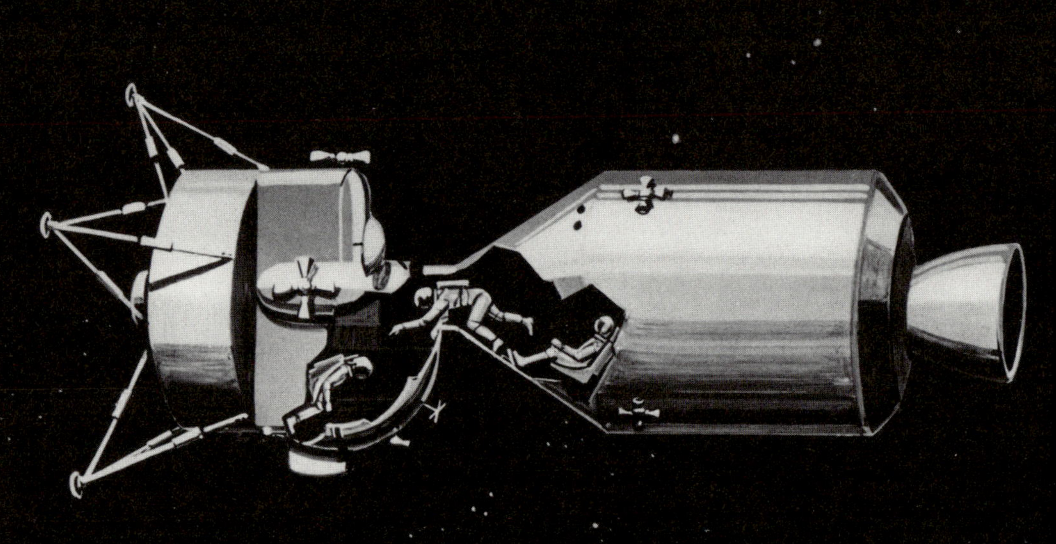

ove left *After docking with lunar module the two craft e separated by spring usters and the spacecraft ves the third stage behind.*

low left *The transfer to lunar module.*

ove *The spacecraft proaches the moon back-rds and retro-fires to enter unar orbit.*

Above *The lunar module descends, while the command service module stays in orbit.*

Above right *Man lands on the moon – an artist's impression.*

Below right *Astronauts set up scientific equipment on the moon – an artist's impression.*

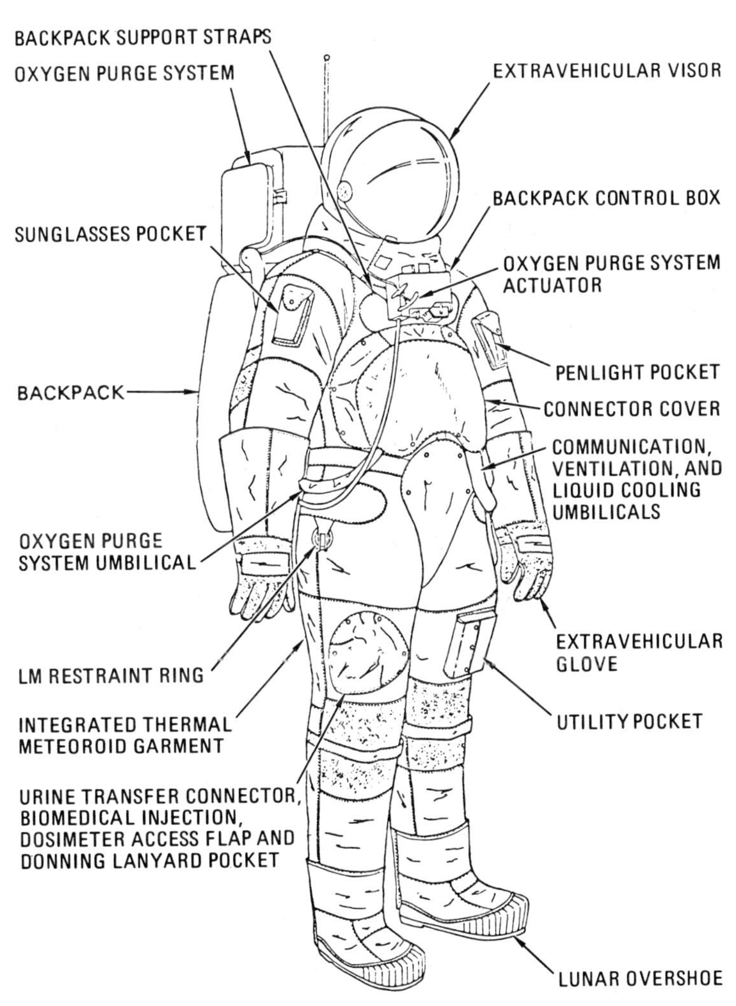

Above *The lunar space-suit.*

Above *The ascent stage of the lunar module is fired to leave the moon.*

Above right *After transferring back to the command service module, the ascent stage of the lunar module is jettisoned and the CSM heads for earth.*

Below right *After separating from the service module shortly before re-entry, the command module is oriented blunt end forward for the aft heat-shield to absorb the 5,000 degree heat as it hits the atmosphere.*

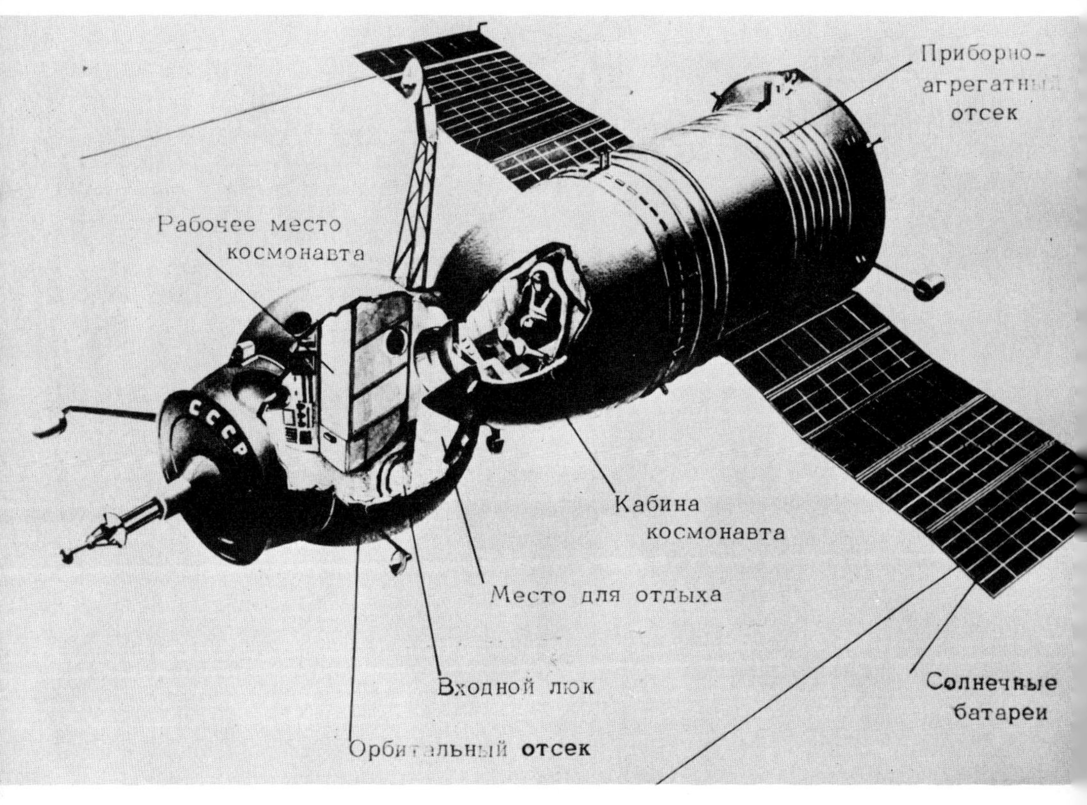

Above *The Soyuz spacecraft. Clockwise: Instrument and systems compartment (probably similar to the American Apollo service module), Solar panels, Cosmonaut cabin, Rest compartment, entrance hatchway, orbital compartment, Cosmonauts' work area.*

1969 The final laps of the race

Without doubt 1969 will go down in history as the year of Apollo. Appropriately the moon landing programme and its spacecraft had been named after one of the busiest and most versatile of the Greek gods. Apollo was the god of light, and twin brother of Artemis, the goddess of the moon. He was the god of music and the father of Orpheus. At his temple in Delphi, he was the god of prophecy. He was also known as the god of poetry, healing, and pastoral pursuits.

On 27 December 1968 Borman, Lovell and Anders, the crew of Apollo 8, safely splashed down in the Pacific at the end of a brilliantly successful mission. They had proved on their Christmas-around-the-moon flight that man could safely leave earth's influence in a spacecraft and return. They had seen with their own eyes what man had never seen before. On the outward flight Colonel Borman had thrilled listeners back on earth as he reported 'I can see at the same time Florida, Africa, Gibraltar, Cuba, and Central America all the way down to Argentina and Chile.' Captain Lovell gave the first close-up description of the moon's surface: 'The moon is essentially grey. No colour. Looks like plaster of Paris.' They saw 'earth-rise' and took many remarkable photographs of it. They also relayed man's first television show from outer space back to earth: it was one of the most moving moments in the history of Apollo as, on Christmas Eve, the camera was switched on to reveal the desolate lunar landscape below. Major Anders was the first to speak, 'We are now approaching the lunar sunrise and for all the people back on earth the crew of Apollo 8 have a message that we would like to send to you.'

Then, in turn, Anders, Lovell and Borman read the ten verses of Genesis, Chapter One.

> In the beginning God created the heaven and the earth. And the earth was without form, and void; and darkness was upon the face of the deep. . . .

Apollo 7 with its earth orbital flight, and now Apollo 8 with its moon orbital success had paved the way for the next stage – the proving trials of the lunar module. A mood of jubilant confidence spread through NASA – the moon-race seemed as good as won, although many still found it hard to believe that all the well-known hazards and risks of space flight seemed to be being overcome with such speed and apparent ease.

Then, at the very end of December 1968 an announcement from Russia caused concern among space experts in the west – the Russians revealed an experiment which could mean they were about to begin a programme of distant interplantary flights. For twelve months, from 5 November 1967 to 5 November 1968 three Soviet scientists had lived in a small airtight chamber only connected with the outside world by telephone and television. They tested a series of possible spacecraft life-support systems. At the same time each man was closely studied to evaluate the psychological and physiological consequences of such an exercise. Their drinking water was regenerated from their own urine, and they supplemented a dehydrated food diet with watercress and other plants grown in a miniature greenhouse.

The success of the experiment meant that really long duration space flights were feasible for man – the biggest difficulty being not the scientific problems involved, but human nature itself. Tempers became frayed very quickly, and one of the scientists commented, 'We have now perhaps become more tolerant of the people around us'.

In October 1968 a leading Soviet spokesman on space, Leonid Sedov, had said, 'It is not possible to be positive about plans for an eventual lunar landing'. He had pointedly added, 'It is also possible that it does not have first priority'. Were the Russians preparing for a trip to Mars? There was much speculation in the space lobbies.

Then, a week after Borman, Lovell and Anders had received a tumultuous ticker-tape reception in New York, news came of a Russian achievement in space which, according to Sir Bernard Lovell, gave the Soviet Union a four-year lead over the United States on the techniques of assembling space equipment in orbit.

On 16 January 1969 Soyuz 4 docked in space with Soyuz 5. It was the first docking ever between two manned craft. Soyuz 4 approached its target under automatic pilot to within 100 yards; then Colonel Vladimir Shatalov, its one-man crew, took over manual control to bring the two spacecraft slowly together. As the two vehicles locked together Shatalov cried, 'I've been hunting all over for you!' – and one of Soyuz 5's crew replied 'Yes, and now you've raped us!'

Two of the Soyuz 5 crew transferred to Soyuz 4 while the docked craft continued to orbit for over four hours. The Soviet press hailed the achievement as a break-through which would lead to huge orbital space stations, the starting points for further journeys into the depths of outer space. Sir Bernard Lovell was reported as saying that if all went well with the Soviet programme, 'my guess is that the Americans will be looking rather small fry by the middle 1970's'.

After the safe landing of the four cosmonauts, *Pravda* printed a seven-point programme for Russian space stations, and it became clear that, if the USSR was interested in sending men to the moon and beyond, they would use such space stations as staging points on their journey. Western viewers later saw television recordings from the interiors of the Soyuz spacecraft, revealing them to be much larger and more spacious than their American counterparts.

Prestige has regularly been quoted as a major stimulus of the space race, with Russia and America competing for headlines. The first weeks of 1969 had an element of farce as the two giants tried to outdo one another, using Press releases for clubs. The United States struck the first blow with the Apollo 8 success. On 6 and 8 January the USSR replied, announcing the successful launch of Venus 5, an unmanned probe continuing the research of Venus 4 (1967). On 9 January the US announced their moon-landing team for Apollo 11. Two days later the

USSR countered by revealing the launch of another vehicle going to Venus, Venus 6 – the two craft would jointly carry out a research programme into the atmosphere of the planet in different regions. The Americans replied with more releases about Apollo 9. Then came Soyuz 4 and 5, and releases about American intentions after the moon landing. Surprisingly perhaps, the public seemed to have an almost inexhaustible appetite for space news, although the continuing issues of Vietnam and Czechoslovakia caused some space-weary critics to suggest that the two great powers had sufficient trouble on earth without looking for it in space.

In March the United States began the last lap of the moon race when James McDivitt and 'Rusty' Schweickart crawled through the hatch into the cramped lunar landing vehicle to become the first American astronauts to transfer from one spacecraft to another in orbit. This was the earth orbital flight of Spider (LM) and Gumdrop (CM). By proving the spaceworthiness of the lunar module it paved the way for a lunar landing in July. But before that flight could take place, NASA insisted on a full-scale dress rehearsal, trying the LM out within 8 miles of the lunar surface. Only after that would it consider making the final step.

However on 18 May, as Stafford, Young and Cernan lifted off from Cape Kennedy for a 'fantastic ride', but probably the most frustrating of all space missions ever (having been expressly forbidden from even attempting a lunar landing), the Russian headlines were filled with the latest news from their Venus probes which had just successfully entered the Venusian atmosphere. A Russian designer was quoted in *Pravda* as saying that further Soviet plans were concentrated on unmanned flight, eliminating the need for oxygen and food. He also pointed out that it would take a spacecraft three years to reach Mars, and that man was not needed to control flights over great distances. From that comment many now inferred that Russia had withdrawn from the race to the moon and were concentrating on unmanned projects. The perennial argument about the merits and dangers of manned spaceflight reappeared in countless newspaper columns. 'Should a man, however courageous he is, be allowed to risk his life when a machine can do the same job?' NASA however was not going to be dissuaded from continuing manned flights now that its goal was so close. They replied with an argument put forward several years before by Scott Crosfield, a test-pilot of the North American X-15. On being asked why send a man to the moon he commented, 'Where else would you get a non-linear computer weighing only 160 lb., having a billion binary decision elements that can be mass-produced by unskilled labour?'

Astronaut safety was not the only subject to come under fire during the successful flight of Apollo 10. University and medical research teams were beginning to protest at possible 'inadequacies' in NASA's planned isolation for the Apollo 11 astronauts when they came back from the moon. NASA had rejected a report which it itself had commissioned, a confidential 600-page treatise prepared by the Baylor University College of Medicine in Texas. The report contained opinions from more than 100 bacteriologists, biochemists, and physiologists. It suggested procedures to safeguard against any infections or toxic agents which might return from the moon with the astronauts or their samples. It proposed a 60-day quarantine for both the crew and everything in their spacecraft. However for the sake of the astronauts' comfort NASA decided to relax its precautions against the spread of alien organisms; the quarantine period was set as three weeks.

Originally the plan had been for the Apollo 11 crew to remain inside their command module until it was safely aboard the recovery vessel, USS Hornet, and within the sealed confines of an isolation area. In the end, NASA decided to drop these precautions, and many others which had also been recommended by medical authorities on the grounds that they were impractical. As the moon is constantly bombarded by cosmic radiation, and exposed to enormous extremes of temperature, it seemed unlikely that any form of life as we know it could possibly exist there. However many scientists pointed out that one of the main reasons in going to the moon was to investigate the possibility of other forms of life. It may be many years before we can evaluate the wisdom of NASA's decision to consider extra precautions 'impractical'. They were probably right, but the lunar landing was a step into the unknown, and if they were wrong, if some dangerous organism which returned with Apollo 11 eventually finds our atmosphere a good place to live, the consequences could be appalling. In that context Frank Borman's remark in January 1969 seems poignantly ironic – 'I look on the Apollo programme as technical life assurance.'

On 9 January 1969, after months of speculation, NASA announced the crew for Apollo 11. As Commander of the flight, perhaps to emphasise its non-military nature, they picked a civilian: 38-year-old Neil A. Armstrong. He was to become the first Man on the Moon.

To meet him, he seems the epitome of the average American, a very ordinary, regular guy. His conversation, however, centres on only one thing – space. Of all the astronauts he is probably the one to have escaped from the greatest number of nasty situations. In 1968 he had an extremely narrow squeak when he left it to the very last second before baling out of a lunar module trainer seconds before it crashed into the ground. He flew in Gemini 8, the only American manned space flight so far which has had to be brought back early due to an emergency. His coolness and skill on that mission received considerable praise and must have been one of the factors which clinched his selection as America's Moon-man. As a test-pilot, before being chosen as an astronaut in 1962, he flew the experimental X-15 to over 200,000 feet and a speed of 4,000 mph; he survived many near misses in research aircraft.

In the Korean War, Armstrong flew 78 combat missions for

the Navy, winning three medals, and once escaping from behind the enemy lines after being shot down. Before his flight to the moon he explained his attitude to Apollo: 'I have been in relatively high risk businesses all my adult life. Few of the others, however, had the possibility of direct gains in knowledge which this one has. I have confidence in the equipment, the planning, and the training. I suspect that on a risk-gain ratio, this project would compare very, very favourably with those to which I have been accustomed in the past twenty years.' It should come as no surprise to learn that his hobby is gliding (for which he is a Fédération Aéronautique Internationale gold badge holder). Even as a child he read aviation magazines in preference to comics, and before he was old enough to drive a car he had been awarded a flying licence.

Today he is often described by people in NASA as 'probably the best jet test pilot in the world' and even though he is a civilian, 'He's more Air Force than the Air Force'. He fits like a glove the NASA image of what every astronaut should be – married, two children, 4,000 hours of logged flying time, and Bachelor of Science in Aeronautical Engineering. Virtually the only criticism that anyone made of this remarkable, utterly meticulous man was his cryptic terseness. Some journalists expected the astronauts to wax loquaciously poetic during the flight – they were disappointed. Armstrong is the kind of man who rarely speaks at all, and when he does, it is brief and to the point. Many suspected after the mission that even his 'one small step for a man' speech had been forced on him by the NASA Public Relations team.

The other man selected to fly down to the lunar surface as lunar module pilot only lost the chance of first-footing the moon because of a NASA change in plan. Originally NASA had decided that Edwin ('Buzz') E. Aldrin Junior, a USAF Colonel, should leave the LM first, and only later was the privilege given to the flight commander himself. Surprisingly, although he was called the lunar module pilot, Aldrin's main job was to check the systems and monitor Armstrong who did most of the actual flying.

'Buzz' Aldrin, a Doctor of Science in Astronautics, was the perfectionist of the Apollo 11 crew. He has been called 'the astronaut's astronaut', and the people who trained him complained that he actually drove *them* too hard – 'the most pernickety astronaut there is!' Superficially he is similar to Armstrong in character, using few words and displaying little humour. He is reported never to make mistakes – 'a good man to have around on a voyage into the unknown!' Aldrin's father, himself a distinguished aviator in the 1920's, was the man who, by a one in a million coincidence, introduced Charles Lindbergh to Dr Robert Goddard in the early days of rocketry.

To get his doctorate, 'Buzz' Aldrin wrote a thesis which is today recognised as a major contribution to the science of docking in space; he proved his theories on the last Gemini flight in 1966, setting a record at the same time for space-walking – five and a half hours. On that flight of Gemini he gained quite a reputation for his suspicion of computers – several times Mission Control overheard him double-checking the computers with his own slide-rule. He has been reported as describing himself as 'a sort of mechanical man'. Others have called him 'the best scientific mind we have sent into space'.

The craft Armstrong and Aldrin were eventually to fly was the world's first true spacecraft. It had only been flown twice before, and the crews of Apollo 9 and 10 were not flattering when they described it: 'A tin foil joke . . . like flying in a bathtub with somebody banging on the outside!' None of it returns to earth, and to fly it one must stand at the controls – there's no room to sit and nothing to sit on. It was aptly dubbed 'Spider' on Apollo 9: the LM's clumsy shape is purely functional, and, as it was never intended to fly in air, it is certainly the most unaerodynamic shape of the space age. However, thanks to Armstrong and Aldrin, it did its job.

The third member of the team was probably the only man in NASA not to see either the moon landing or Armstrong's first steps on the lunar surface. He couldn't – there was no television monitor aboard the command module, which he was piloting in lunar orbit at the time. 38-year-old Colonel Michael Collins (USAF) also was faced with the possibility of the most unpleasant task in space history. If anything went seriously wrong aboard the lunar landing vehicle before it docked with the CM for the three astronauts to return to earth, he would have to leave his two colleagues behind. As he said, 'If they have difficulty on the surface of the moon, there's nothing I can do about it . . . I don't think that will happen and if it did I would do everything I could to help them, but they know, and I know, and Mission Control knows that there are certain categories of malfunctions where I just simply light the motor and come home without them.'

Like his companions aboard the spacecraft he had considerable test-pilot experience, but in character he is a very different sort of man. Of Irish descent, born in Italy, he comes of a military family. With a general for a father, an Army Chief of Staff for an uncle, and a brigadier as brother, it was almost inevitable that Mike Collins himself should be a West Point man. However he is not obsessed with space – he's just as keen on increasing his considerable knowledge of wine, and pruning the roses in his backyard.

As pilot on the three-day, 44-orbit, Gemini 10 mission, in 1966, he shared the accomplishments of that record-breaking flight with astronaut John Young. The flight gave him docking experience (with two Agenas), and his skilful performance during the two EVA's contributed greatly to NASA's manned space flight knowledge. He's also a very fortunate man: he was dropped from the Apollo 8 mission because of a loose disc in his neck which was pressing on the spinal chord. He is the astronauts' hand-ball champion, and in 1968 his game got steadily worse, his reflexes began to go, and occasionally he fell down. Doctors told him that if he came through two dangerous operations he might become fit again. He opted for surgery – 'I had no alternative' – and his optimism paid off.

To put those three men around the moon involved the biggest communications exercise in history. It necessitated a two-million-mile hook-up of land-lines, undersea cables, radio circuits and communication satellites. The Manned Space Flight Network depended on both a world-wide chain of 30-ft radio antennas (for tracking, command and communication) for the earth orbit stages of the mission, plus three 85-ft diameter antennas (in Spain, California, and Australia, providing 120-degree coverage) for all the stages more than 10,000 miles from earth. For Apollo 11 the network used 17 ground stations, 4 ships, covering areas beyond the range of land stations, and eight Boeing 707 jets covering the gaps between land stations and ships – particularly in the Pacific area between Australia and Hawaii. This network was not merely for human voice communications: it was also essential to enable the computers both on the ground and in the spacecraft to exchange data at the incredible speed of 2,400 bits of information every second.

Without such a network Apollo 11 would have been impossible. But even with it, the flight was far more dangerous, with more 'Life Critical Points' than many realised. However, it is highly unlikely that Armstrong, Aldrin and Collins carried suicide pills. On being asked what his colleagues would do if they found themselves doomed to an eternal orbit of the sun, astronaut Schmitt (a geologist destined to fly on a future Apollo mission) told me he felt they would just continue space studies, relaying information to earth until their life support systems finally ran out.

In the bid to land on the moon, despite all the experience gained from previous missions, the astronauts faced several truly 'Life Critical' moments. The first was on the pad: the tragic oxygen flash-fire of 1967 was a clear reminder of what could happen. Then, during lift-off if two or more stages of the rocket failed the great Saturn V would topple over. In an 'Abort' at that moment they would have had to rely on the launch-escape-tower system to rocket them away to safety.

If the third stage of the Saturn had failed to burn for precisely the correct duration during the translunar insertion, they could have missed the moon altogether and become trapped in an eternal solar orbit. On approaching the moon the tricky manoeuvre of lunar orbit insertion is carried out behind the moon, out of touch with earth. If the engine had overfired, the craft would have slowed excessively and crashed on the moon.

Any mistake during the lunar module's descent could have been fatal. The LM had to land squarely: re-launch was impossible if it tilted more than 12 degrees from the vertical.

Also, as no one really knew what the lunar soil was going to be like, this was to be the most dangerous moment of all, for a seemingly good surface might have suddenly caved in. Armstrong's steps into the unknown were a calculated risk, but even within the comparative safety of the lunar module, a sudden solar flare would have been deadly: neither their space-suits nor the lunar module provided adequate protection against unexpected increases in radiation, and their only hope of survival would have been a quick return to the protective shielding of the orbiting command module – assuming sufficient warning had been given for them to reach it in time. The lunar lift-off, rendezvous, docking, transearth insertion and even re-entry into earth atmosphere all provided extra opportunities for something to go fatally wrong. The astronauts must also have been very conscious of the discovery, after the 1967 holocaust that '113 significant engineering orders' were shown never to have been carried out. Recent successes encouraged the hope that standards had improved – but in such a vast project with so many thousands intimately involved there was always the possibility of one little human error.

Wherever possible on the Apollo project, 'Redundancy' was built into the systems involved: that meant that if any component went wrong in the rocket, spacecraft, or at ground-control, there was another component or back-up system which could take over its function. However several vital systems could not be duplicated (e.g. there was only one engine in the LM to get them off the moon, only one engine behind the CSM to get them back from the moon). Today we know nothing did go seriously wrong – but before the flight the astronauts knew only too well that their lives depended on the perfect functioning of an upsettingly large number of components.

Launch day had been chosen as 16 July, the exact timing being decided by the fact that it gave the longest 'launch window' for a launch from Cape Kennedy to the moon, at that time of year. As the great day got closer, 'moon fever gripped the world' – or at least moon fever gripped the world as far as the Press were concerned. Almost every newspaper, periodical, and magazine on earth was packed with information about the forthcoming flight. Not all of it however was encouraging for NASA's future after the moon landing. Ironically the brother of the man who committed the US to the moon trip was suggesting a cutback in space expenditure. The crueller irony was that his later car crash may have put his career in jeopardy, detracting from the validity of his remarks before Apollo 11. 'A substantial portion of the space budget can be diverted to the pressing problems here at home.'

'We should develop a plan for an orderly, programmed exploration of outer space, but we no longer need an accelerated programme. We need not try to get to Mars or Venus merely because the Russians might get there first . . . I am for the space programme, but I want to see it in its right priority.'

Senator Edward Kennedy's point about the futility of space rivalry was expanded in a leading article by Bertrand Russell, published in the London *Times* on the eve of the launch of Apollo 11. Under the title **Why man should keep away from the moon,** Lord Russell wrote '. . . Unfortunately the spirit of ruthless competition has infected projects for reaching the moon. These projects were not considered in a spirit of scientific detachment, or as redounding to the credit of the human race. They were regarded, instead, as an opportunity for a race between great rival powers. It was felt that the important thing was not that the moon should be reached, but that it should be reached by our side (whichever that may be) sooner than by the other. This is paltry, and makes the whole enterprise one in which it is difficult for sane men to see much of value.'

On the subject of the concern of scientists at polluting the lunar atmosphere he pointed out that we would do well to remember that we do the same to our own earth environment all the time – 'To my mind there is a lack of proportion in this meticulous care for the moon combined with wanton damage to our own planet as the bearer of life.'

The paltry rivalry of which Lord Russell wrote was only too apparent a couple of days before his article appeared. On 13 July, three days before the scheduled Apollo 11 lift-off, the Russians launched Luna 15. Alexei Leonov, the Soviet cosmonaut, had already told correspondents in Moscow that the Soviet Union would display a sample from the moon at Expo 70, and indicated that it might be retrieved before the American landing. The flight of Luna 15, therefore, surprised no one. Most experts thought it might well be a moon-scoop device designed to return automatically to earth: it wasn't, but as a headline scoop for the USSR it proved highly effective for several days.

On 16 July 1969 a million people converged on Cape Kennedy to see 'The Big One': that's how Rocco Petrone, the Director of Apollo 11 launch operations, described it, 'the one we've been working on for eight years'. Amongst the crowd in the VIP stand were Hermann Oberth, the German pioneer, and Charles Lindbergh, the first man to make a solo flight across the Atlantic, and the man who gave Dr Robert Goddard such invaluable assistance in the early days of rocketry. Ex-President Johnson was there. Ironically it was rumoured that Mr Johnson was justifiably piqued when, by some NASA oversight, he was amongst the last to receive an invitation to watch the moon launch.

Thousands camped out all night in cars, tents, and caravans, and after dawn film helicopters revealed miles of messages scrawled along the Florida beaches saying 'Good Luck Apollo 11'. 3,400 journalists turned up to report the event to every corner of the globe, including 111 from Japan alone.

Around the world an estimated 1,000,000,000 watched on television as the countdown reached its final moments. Exactly on schedule, at 8·9 seconds before 9.32 a.m. in Florida, the five engines of the first stage of the Saturn V were ignited: with a shattering roar the great moon rocket built up its 7,600,000 lb. thrust and at 9.32 precisely a relieved but proud voice from Mission Control told the world 'We have lift-off!' At that moment the service arms of the mobile launch tower retracted, and a deluge of 50,000 gallons of water a minute began to cool the tower to stop it melting under the scorching blast of the rocket flame. Below the rocket a specially built flame deflector in the flame trench received the full impact of the heat of lift-off. Its refractory concrete and volcanic ash surface turned to glass, despite a colossal water dousing.

Very slowly at first, the giant rocket began to rise into the sky. For a couple of seconds it was obscured by the smoke and dust of its own exhaust. Then, as the sound waves began to rattle the windows of the firing rooms, the Saturn V re-emerged, a majestic white column silhouetted against the blue sky. The ground shook around the Press Stand as the rocket, as tall as St Paul's Cathedral, picked up speed. Within seconds one could see it begin to tilt over towards the East as it accelerated towards orbit. Piercing through layers of light cloud, the huge machine cast a strange shadow which sped across the sky like a mysterious ghost ship.

Pictures transmitted from a camera aboard a high-flying aircraft enabled television viewers to see the rocket as the spectacular first-stage separation took place. Seconds later the spent first-stage dropped into the Atlantic some 340 nautical miles down-range. NASA experts were already saying that it was the smoothest Saturn launch to date. Control of the mission had now passed from the firing rooms at Cape Kennedy to Mission Control in Houston, Texas. As the control team monitored their display boards, they saw the final steps towards the climax of the Apollo programme being achieved with almost unbelievably meticulous precision. Only one thing was causing concern: Luna 15. Were the Russians going to win the moon-race after all? Was Luna 15 a manned mission? The radio telescope at Jodrell Bank had tracked the Russian spacecraft and revealed it to be travelling much more slowly than any earlier flight in the Lunik series. Most American experts were now convinced that the Russians were attempting the experiment they had feared – the automatic lunar landing, followed by the 'moon-scoop' and return to earth with samples before Apollo 11.

Twenty minutes after lift-off, and 2,300 nautical miles down-range the second stage of the moon-rocket plunged into the Atlantic west of Africa. Eight minutes before that splash the third stage had successfully pushed the Apollo spacecraft a further 2,500 nautical miles, putting it into earth orbit at precisely 25,567 feet per second. With typical mastery of the under-statement Neil Armstrong, the flight commander, reported back, 'We have no complaint with any of the three stages on that ride. It was beautiful.'

Midway over the Pacific, and half way round the earth for the second time, the S-IVB third-stage engine was re-ignited.

Mission Control radioed 'You're looking good, you're go!' Speed increased to 24,545 mph and the spacecraft began its long journey to the moon, the translunar insertion (TLI).

Up to four midcourse correction burns had been planned during the translunar coast phase of the mission. However, the accuracy of the trajectory resulting from the TLI manoeuvre was such that only one correction was needed; the crew were able to relax as their craft hurtled forward, rotating slowly on its axis in the 'barbecue' mode, maintaining a steady temperature under the constant solar grill.

Three days after lift-off the Apollo spacecraft was scheduled to go into lunar orbit. The day before, the astronauts delighted television viewers back on earth by entering the lunar module earlier than expected, and transmitting a guided tour of their cramped quarters in vivid colour. On both Apollos 9 and 10 the astronauts had experienced some difficulty in getting through the narrow hatch between the lunar and command modules. On Apollo 11, however, the crew seemed to overcome the problem with ease, and the wide-angle lens of their camera relayed pictures which made their new confined home look almost spacious.

NASA, while pleased with the success of this early part of the mission, was now getting very worried about Luna 15. There seemed little chance of a collision between Apollo 11 and the Russian spacecraft, but the Americans wanted reassurance. Colonel Frank Borman, the commander of Apollo 8 had just returned from his visit to the Soviet Union; as Ambassador Extraordinary for American space travel he telephoned the President of the Soviet Academy of Sciences, Academician M. V. Keldysh. Keldysh promised him that Luna 15 would not intersect the published trajectory of Apollo 11 at any point, and that he would be kept informed of any changes. Confusion grew again in the Press when a Tass commentary mentioned 'space stations landing on the lunar surface'. However space reporters suggested that 'a degree of cooperation on space exploration had been achieved by both countries which was greater than either was prepared to divulge.'

By the third day of the mission Armstrong, Aldrin and Collins had settled down to the routine of space flight, and compared with the relatively primitive conditions aboard the Mercury capsules, Apollo travel was space luxury at its best. Each astronaut had his own menu for the day, with extra food available from a 'snack pantry' to avoid breaking into the daily food packages or 'robbing' a regular meal deep in the storage box. On Day 3 Neil Armstrong's menu was as follows:

Breakfast: Peaches, Bacon Squares (3), Apricot Cereal Cubes (4), Grape Drink, and Orange Drink.

Lunch: Cream of Chicken Soup, Turkey and Gravy, Cheese Cracker Cubes (6), Chocolate Cubes (6), and Pineapple-Grapefruit Drink.

Dinner: Tuna Salad, Chicken Stew, Butterscotch Pudding, Cocoa, and Grapefruit Drink.

The astronauts had over 70 items to choose from in a list of 'freeze-dried rehydratable, wet-pack, and spoon-bowl foods'. Later in the mission one saw them on television as they kneaded their food in plastic containers. The 'snack pantry' contained 75 drinks, ranging from coffee to grape punch (presumably non-alcoholic), plus over 100 food items including twelve strawberry cubes, three sausage patties, six portions of spaghetti and meat sauce, and twelve peanut cubes. In spite of the danger from floating crumbs they carried 72 slices of rye and white bread, on which they applied sandwich spreads ranging from tuna salad to Cheddar cheese. Each man-meal package also contained a small wet-wipe cleansing towel, and under the command module pilot's couch were stowed 21 additional dry towels in case the crew found eating in weightlessness a messy business.

Previous Apollo missions were bedevilled with ill-health – everything from nausea to the common cold. On Apollo 11 the NASA medical advisors were taking no chances. The spacecraft was equipped with a medical kit which, although small in size (5×5×8 inches), was adequate to cope with the needs of the most hypochondriac astronaut. The box was stowed beside the lunar module pilot's couch. It contained three motion sickness injectors, three pain suppression injectors, one two-ounce bottle of first-aid ointment, two one-ounce bottles of eye drops, three nasal sprays, two compress bandages, 12 adhesive bandages, one oval thermometer, and four spare crew biomedical harnesses (which relay astronaut health information back to earth). Also in the kit were nearly 300 pills: 60 antibiotic, 12 anti-nausea, 12 stimulant, 18 painkiller, 60 decongestant, 24 diarrhoea, 72 aspirin, and 21 sleeping. In addition a similar but smaller medical kit was carried to equip the lunar module against astronaut illness. 'They think of everything!' – I heard several astronauts refer gratefully to the NASA team in such terms.

The record of accuracy in splashdown locations has steadily improved – television viewers today expect cameras to find and show the command module even before its parachutes open prior to splashdown, after re-entry into the earth's atmosphere. Such feats of navigation would have been incredible only a few years ago: today they are an accepted commonplace. But, just in case things did go wrong, and the returning Apollo astronauts landed in some very unexpected part of the world, they flew prepared. Above the lunar module pilot's couch is the right-hand forward equipment bay: in it are stowed two rucksacks containing survival gear. One contains an inflatable life-raft, with a sea anchor, sea dye markers, sun bonnets, etc. The other holds a radio beacon, water containers, a desalter kit, sunglasses, sun lotion, and the inevitable survival knife. The kit is designed to 'provide a 48-hour postlanding (water or land) survival capability for three crewmen between 40 degrees North and South latitudes'.

'All systems go' radioed the Apollo 11 spacecraft. 'Now going round the moon. See you on the other side. Everything looks OK.' A second later, and a hiss of static indicated that radio contact with the command module was lost. Ten minutes later, behind the moon, and three days, three hours, fifty-four minutes and twenty-eight seconds after lift-off, the first lunar insertion burn was made. At that moment Apollo 11 was just 80 miles from the lunar surface. The single main engine of the spacecraft had to brake Apollo to about 3,600 mph: any more, and Apollo 11 would crash into the moon; any less and the astronauts would sail beyond the moon into a long trajectory which would eventually return them to earth.

It was at this time that a very slight fuel leak developed. Only when Apollo emerged back round the other side of the moon could Mission Control learn of its existence: fortunately it was not serious, and it was part of a redundant system (i.e. a system for which a back-up substitute was available).

The astronauts were now able to enjoy their first good view of the moon. As they flew over the crater of Aristarchus, Armstrong commented: 'It's an area that's considerably more illuminated than the surroundings. It seems to have a slight amount of fluorescence. The area in the crater is quite bright.' Many astronomers have observed bright spots on the moon before – some consider they are evidence of volcanic activity beneath the lunar surface. Armstrong's observation was even more interesting because several reports had been received in Houston from astronomers of bright spots in the Aristarchus area during the outward flight of Apollo 11.

Five hours later, and again behind the moon, the elliptical orbit was adjusted by a second lunar insertion burn. By the time the lunar module was preparing to return and rendezvous after the landing mission, this new orbit was designed to balance itself into a circular one, 60 nautical miles from the surface.

The testing time for Apollo had arrived. The world was now becoming familiar with the new names for the Apollo 11 spacecraft: the lunar module was called 'Eagle', and the command module, 'Columbia' – an intentional echo of Verne's prophetic 'Columbiad'. Armstrong and Aldrin now had to leave Collins and make the historic descent to the lunar surface. During the thirteenth orbit of the moon the docking collar which locked Eagle to Columbia was released. Then, as the two craft regained contact with earth, Armstrong gave the good news: 'The Eagle has wings.' Collins fired the thrusters on Columbia to pull ahead of Eagle. He said, 'You've got a fine looking flying machine there, Eagle, despite the fact you're upside down.'

Collins then watched closely as Aldrin slowly rotated Eagle: one of Collins' last jobs, before beginning his lonely vigil aboard Columbia, was to make a visual check on the LM's landing legs.

Gently at first, with 10% throttle for 15 seconds, then, increasing power to 40% throttle, Eagle's descent to the moon began. NASA, after months of deliberation, had finally chosen a seemingly flat area on the Sea of Tranquility as the landing site. Both the approach and landing phases of the operation were planned to allow Eagle's pilot to take over manual control: if he disagreed with the way the radar-guided computer was flying his craft, he could take command himself – and in the event this precaution proved to be one of NASA's wisest decisions.

'Looks real good,' radioed Eagle.

'Yes, everything is looking good to us,' replied Mission Control. 'Continue your powered descent.'

'We have the moon right out of our window. Even better than the simulator.' But the particular part of the moon they could see struck Armstrong as far too rocky to risk a landing. It was a crater surrounded by jagged boulders, 10 feet across. He took control and overflew the rocks, to come down about four miles from the planned landing site. At three and a half feet per second Eagle sank towards the surface. Five feet before the landing pads touched the moon the four sensitive probes below them jabbed the surface, and signalled 'lunar contact' on the control panel. A second later the descent engine was shut down and Eagle landed. The historic touchdown was at 2.17 p.m. (Florida time), a mere 81 seconds ahead of flight schedule.

Colonel Aldrin reported 'Tranquility Base – The Eagle has landed.' The millions of sighs of relief on earth were summed up by a comment from Houston: 'We got a bunch of guys on the ground about to turn blue. We're breathing again!' So did Armstrong – on landing his pulse raced at 156 beats a minute, but now it relaxed to a cool 90. Eagle was resting on the moon at an angle of 4½ degrees – another cause for relief, for it was well within the limits set for a safe take-off.

Aldrin began to describe his lunar surroundings. '. . . it looks like a collection of every variety of shape, angularity, granularity, a collection of just about every kind of rock. Colour depends on what angle you're looking at. Rocks and boulders look as though they're going to have some interesting colours.' Armstrong added, 'I'd say the colour of the local surface is very comparable to that we observed from orbit at this sun angle. It's pretty much without colour. It's grey, a chalky grey as you look at the sun line, and a darker, ashen grey as you look at 90 degrees of the sun.'

Having overflown their intended landing site, Armstrong was not too sure exactly where they were on the lunar surface. He got the comforting assurance from earth: 'Roger Tranquility: no sweat, we'll figure it out!'

'Don't forget one up here' – Collins in Columbia asked Eagle to switch to open antenna so that he could hear what was going on below – 'Or I'll miss all the action.'

'The moment of touchdown was one of the moments of greatest drama in the history of man. The success in this part of the enterprise opens the most enormous opportunities for the future exploration of the Universe'. With these words the astronomer Sir Bernard Lovell greeted the news of the landing. In America President Nixon said the achievement would 'stand through the centuries as one supreme in human experience and profound in its meaning for generations to come'. The Pope watched the flight on colour television. He made an unscheduled speech: 'Glory to God in the highest and peace on earth to men of good will! We, humble representatives of that Christ, who, coming among us from the abyss of divinity, has made to resound in the heavens this blessed voice, today we make an echo, repeating it in celebration on the part of the whole terrestrial globe, with no more unsurpassable bounds of human existence, but openness to the expanse of endless space and a new destiny.' It was Buzz Aldrin's wife, however, who summed up most people's feelings as she watched the moment when man first began his colonisation of other worlds: she said, 'I just can't believe it!'

Luna 15 had now moved closer to the lunar surface, and was orbiting barely ten miles above the moon. At that altitude the slightest error in its trajectory would cause it to crash into one of the higher peaks of the moon mountains. Moscow radio said nothing, but announced the American landing in its 11.30 news bulletin.

Before stepping out of Eagle to make history, Armstrong and Aldrin now had to check all the systems aboard the Ascent stage of the LM in case they had to make an emergency escape from Tranquility Base. Then they consumed man's first meal on the moon. It consisted of four bacon squares, three sugar cookie cubes each, peaches, a pineapple-grapefruit drink, and coffee. As we saw in Chapter 12, they were now scheduled for a sleep period, but the excitement was presumably too much, for they soon requested permission from Houston to begin their exploration of the moon surface. It was granted. After eating, and a short rest, the astronauts donned their moon-suits, the extravehicular mobility units (EMU) which can keep a man alive for up to four hours on the moon without returning to the lunar module.

Inside the spacecraft, astronauts wear a one-piece constant-wear garment, rather like 'long johns'. Armstrong and Aldrin had to unzip this, and replace it with the liquid cooling garment; it's worn next to the skin, and is made of knitted nylon-spandex, with a network of plastic tubing through which cooling water is circulated. This water comes from the portable life support system (PLSS), a backpack which also supplies oxygen, and contains its own power supply for an independent set of displays, controls, communications and telemetry equipment.

There was no room for a lavatory in either Eagle or Columbia, with the result that personal hygiene was always an uncomfortable problem for the astronauts. Under even the liquid cooling garment they had to wear the space version of baby's diapers – a pair of highly absorbent elasticised briefs. After excretion the solid body wastes were collected in plastic defecation bags. These bags were then sealed and stowed in empty food containers. Urine was collected in one-litre capacity bags worn flat on the lower abdomen. The urine from these collection devices was stored aboard the LM to be dumped overboard later through a special valve in Columbia. On an earlier mission, one astronaut, using a space version of the mariner's sextant, tried to plot his position by some 'stars'. Only when the computer disagreed with his findings did he realise that the 'stars' were urine droplets, floating beside his capsule.

Having donned the liquid cooling garment to protect against the extremes of temperature on the moon (one astronaut described it as 'putting on a cold shower'), Armstrong and Aldrin then had to struggle into their Integrated Thermal Micrometeoroid Garments. They are the most expensive suits ever made, costing $100,000 each. Lined with neoprene coated nylon, seven layers of insulation defend the astronaut against heat and cold, whilst an outer layer of Teflon-coated Beta fabric should provide ample protection against micro-meteoroids. Every astronaut has to have his own custom-built suit specially tailored for him, and each garment received 500 different inspections before it was passed as space-worthy. Weighing 50 lb., the suit was clumsily heavy on earth, but, as Aldrin was soon to prove, it permitted considerable freedom of movement in the one-sixth earth gravity environment of the moon. However there were snags: an astronaut wearing such a suit not only cannot touch his toes – he cannot even reach lower than 22 inches from the ground. Hand movements for maximum efficiency are limited to approximately waist level. Because of this, samples of moon dust and rock had to be picked up using special scoops with telescopic arms. Even the gloves were tailor-made, and were built up on plaster casts of each astronaut's hand. The finger tips were made of special silicone rubber, giving the crewmen maximum sensitivity.

The space helmets for the moon carried two visors with thermal control and optical coatings on them. The EVA visor was attached over the pressure helmet to provide impact, micro-meteoroid, thermal, ultra-violet, and infra-red light protection while on the moon surface.

Even with the protection of these visors, many experts on human psychology and perception were fearful at how the first man on the moon would react to his first sight of another world, when he was standing on it. To look at the sun for barely a second would cause blindness, and even reflected light could make temporary blindness a hazard. On earth, as Professor Richard Gregory has pointed out, the sight of distant objects implies a certain haze; on the moon the lack of haze caused by the almost complete absence of atmosphere makes the farthest object seem as close as the nearest. This could confuse any moon-walker. Similarly the absence of familiar or regularly shaped features common on earth, could produce uncertainty and systematic errors of scale. A fall, resulting from such an error could rip the space-suit – leaving only 15 seconds of consciousness, and death within 15 minutes as the blood boils in the lunar vacuum.

'Everything is go here. We're just waiting for cabin pressure to blow enough to open the hatch.' Armstrong was preparing to leave the lunar module.

'We see relatively static pressure on your cabin. D'you think you can open the hatch at about ·150 p.s.i. (lb. per square inch)?'

Aldrin replied to Mission Control, 'We're going to try it.' Then Armstrong announced, 'The hatch is coming open.'

Man's first entry into another world was probably the most ungainly movement ever recorded in human history. To leave Eagle, Armstrong had to kneel, and slowly wriggle his way feet-first, face down, out of the hatch. Aldrin, advising him how to move, provided a running commentary: 'O.K. move up there. Towards me! Lean out a little bit. You're lined up nicely. Toward me a little. Down! O.K. you're clear. Roll to the left. You're lined up on the platform. Put your left foot out a little bit. O.K., that's good, hold that! You're not quite squared away – over to the right a little bit. You're even. That's good. Go slowly on the first step. You're doing fine.'

This was the moment when Armstrong pulled the handle on the outside of the lunar module which released the MESA (Module Equipment Stowage Area) and uncovered the television camera on the side of the descent stage of the lunar module. Back on earth a billion viewers saw a blurred, upside-down picture of the leg of Eagle which held the ladder to the surface. Seconds later it was adjusted, and one could see the shadowy outline of Armstrong as he began to climb down. He was scheduled to descend and pause, standing on the pad at the base of the leg, before stepping into the unknown.

'I'm at the foot of the ladder. The LM footpads are only depressed in the surface about one or two inches. Though the surface appears to be very, very fine grained as you get close to it – like powder.' The time was 10.56 p.m. in Florida, Sunday, 20 July 1969.

'O.K. I'm going to step off the LM now. That's one small step for man, one giant leap for mankind. The surface is fine and powdery. I can kick it up loosely with my toe. The dust adheres in fine layers like powdered charcoal.' Looking down he saw the treadmark of his lunar overshoe clearly printed on the moon.

'There seems to be no difficulty in moving around – it's even lots easier than the simulations at one-sixth g that we performed in various simulators on the ground. Essentially no trouble to walk around. The descent engine did not leave a crater of any size. There's about one foot clearance of the ground and we have essentially a very level place here. I can see some evidence of soil contamination from the descent engine but it's a very insignificant amount.' Several scientists on earth disagreed with him. They were concerned about the contamination of both the moon's soil and its atmosphere. A Grumman Corporation research team informed NASA that the LM exhaust could increase the total estimated weight of the lunar atmosphere by as much as 20% – 'an indication that the moon may soon enjoy its share of terrestrial atmospheric pollution problems'. Considering the fact that Eagle's descent engine poured almost nine tons of exhaust products into the rarified lunar atmosphere, the criticism must be taken seriously – although it is difficult to see how man could ever study the moon first-hand without inevitably bringing some pollution with him. The most immediate problem such pollution created was for the scientists at the Lunar Receiving Laboratory in Houston – when they received the first precious lunar samples they knew their analysis would always be open to some doubt; the possibility of sample contamination could never be totally discounted. Retrieving moon rock from deep under the surface will give a better chance of avoiding pollution, and Armstrong must have been thinking of this as he tried to shovel deeper into the soil. Unfortunately his space-suit made digging difficult – 'I'm sure I could push it in farther, but it's hard for me to bend down farther than that.'

It was now time for the second man on the moon to start exploring. Armstrong supervised Aldrin's descent from the LM. 'O.K., you know the difficulties I was having. I'll try to watch your feet from underneath here.' Aldrin replied, 'Just making sure I lock up on the way out!' While Armstrong laughed, Aldrin explained to his television audience, 'That's our home for the next couple of hours and we want to take good care of it.'

From now on, although sensibly cautious, the two astronauts behaved in such a relaxed manner that it was hard to remember that virtually every move they made and every breath they took was a first, part of the climax of man's greatest pioneering exploit. It sounded more like a Sunday afternoon dialogue between two climbers in the English Lake District:

Aldrin: 'Looks good down there! I think I'll do the same.'

Armstrong: 'There you go. There you've got it! That's a good step.'

Aldrin: 'Beautiful, beautiful.'

Armstrong: 'Isn't that something? You did a good job coming down here.'

Aldrin: 'I accept your congratulations.'

The two men then unveiled the plaque on Eagle's leg. 'Here men from the planet earth first set foot on the moon, July 1969 AD. We came in peace for all mankind.' It bore the astronauts' signatures; below their names Mr Nixon had added his own – much to the anger of his political opponents.

The television camera was now moved some 60 feet away from the LM, and after a series of lunar panoramas it showed its audience back on earth a static view of Eagle.

Armstrong left the camera on a tripod. Then, with elephantine grace, he floated back into the picture like some ghostly knight in white armour. Aldrin could just be seen below Eagle, setting up the solar wind experiment. He unfurled a strip of aluminium foil to trap particles of solar wind, and two hours later he rolled it up again to take back to earth for examination.

There was much controversy back on earth before the

mission about whether or not the astronauts should, in true colonising style, plant the Stars and Stripes on the moon. Many Americans felt that it might reduce the triumph of the event by introducing an element of nationalism into the venture. However Congress insisted that the astronauts were right to display their proud patriotism at such a moment. The moon, however, did not give in easily: it took several attempts before the flag-staff could be forced into the soil. Even then, the ultra-rarified atmosphere would not support the colours, and they had to be erected and straightened on a wire stiffener.

Astronaut Bruce McCandless, the Cap Com (the Capsule Communicator who relayed messages to Eagle and Columbia from Houston) told the astronauts that Mr Nixon was on the line.

McCandless: 'Neil and Buzz, the President of the United States is in his office now and would like to say a few words to you.'

Armstrong: 'It would be an honour.'

With the astronauts standing as erectly as their space-suits permitted, the live picture of the President was superimposed on the lunar scene.

Nixon: 'Hello, Neil and Buzz. I'm talking to you by telephone from the Oval Room at the White House, and this certainly has to be the most historic telephone call ever made.' (Mr Nixon later quipped that he made the call 'Collect'.)

'I can't tell you how proud we all are of you. For every American this has to be the proudest day of our lives. And for people all over the world, I am sure they too join with Americans in recognising what an immense feat this is. Because of what you have done the heavens have become a part of man's world. And as you talk to us from the Sea of Tranquility it inspires us to redouble our efforts to bring peace and tranquility to earth.' President Nixon was about to make a world tour including a very controversial visit to Romania. He was not going to lose the diplomatic opportunities which such a world event offered. 'For one priceless moment in the whole history of man, all the people of this earth are truly one: one in their pride in what you have done, and one in our prayers that you will return safely to earth.'

Armstrong's priority task on the moon had been to quickly retrieve a sample of moon dust, in case something unexpected had caused him to leave hurriedly. He had collected this 'contingency sample' using a scoop on the end of an extendible handle. Now, with the sample safely pocketed, the next task was to dig up as much moon rock and soil as time permitted. While Armstrong filled the 'rock boxes' with bulk material, Aldrin was delighting earth viewers with their first moon circus act. In the jargon of the Apollo schedule his activities were called 'EVA and Environmental Evaluation'. Aldrin tried walking, then speeded up and was seen gambolling across the screen, giving the impression of a slow motion athlete: 'You've got to be careful you lean in the direction you want to go, otherwise you seem inebriated,' he joked.

As Aldrin cavorted around, he left footprints which will be preserved for 500,000 years until they are erased by micrometeorite impact – presuming of course that the historic prints are not trampled underfoot by future generations of lunar tourists.

After making an all-round inspection of Eagle, Aldrin brought out the passive seismometer which relayed moonquake information back to earth after their departure. Like any holidaymaker lugging a heavy suitcase, Aldrin staggered away from the LM to position the seismometer sufficiently far from Eagle to avoid it being damaged by the exhaust during lift-off. Even before the astronauts had splashed down after their return flight this seismometer had recorded a sizeable quake. After the vibrations had been relayed to earth seismologists were heatedly arguing whether they were caused by volcanic activity or a meteorite impact.

Meanwhile Armstrong was busy setting up another experiment – the 'Laser Ranging Retro Reflector'. By bouncing a laser beam on to this device and timing precisely how long it took for the beam to return to earth, scientists were at last in a position to measure the distance to the moon within almost a fraction of an inch – an important aid for future navigation. This experiment could well be the forerunner of an important development in lasers. Professor Kopal of Manchester University recently suggested that solar power-plants, deriving their energy from the sun, would be vastly more efficient on the moon than on earth. The power they could produce (at least a thousand times greater than earth's largest hydro-electric dam) could be beamed to receiving stations on earth by lasers, for conversion into whatever type of energy was required. He added one sinister note, 'An intensive laser source on the moon could, in unscrupulous hands, become a weapon whose destructive effects could be comparable with those of hydrogen bombs, and worse; for coherent beams could also be produced of light invisible to the human eye, and the extent of terror produced could be graduated to any shade.' A Moon Death-Ray sounds like science fiction – but so did nuclear power, until Hiroshima.

Finally Aldrin hammered two core samplers five inches below the surface in the hope of retrieving some guaranteeably uncontaminated material. Then, using a Heath Robinson kind of clothes-line conveyor belt they hauled the rock boxes up into the ascent stage of the LM. Finally, to avoid bringing unsealed lunar contamination into Eagle, the two crewmen wiped their boots on the porch before forcing their way back into the LM through the hatch.

There is no space inside the confines of Eagle for sleeping couches, so the two men had to rest strapped to their harness or huddled on the floor. They slept fitfully for about five hours – Armstrong's heartbeat indicated that he barely slept at all. Collins however had a good night's sleep aboard Columbia. On being woken by Houston he commented, 'Not since Adam has man known such solitude as Mike Collins when he passes round the back of the moon.'

At 1.54 p.m. (EDT), right on schedule, the ascent engine of Eagle fired and lifted the tiny craft from its launch pad – formed by the LM descent stage which was left behind on the moon. Compared with the slow majesty of their earlier departure from earth aboard the great Saturn V, the speed of this exit was almost rude. Within seconds they were thousands of feet above the surface.

The two astronauts flew the moon-bug in a standing position, and Aldrin soon reported: 'Man, that's impressive looking isn't it? We are going right down US One' (a major American highway, and the astronauts' name for a prominent crack in the lunar surface). 'We're going right down the throat – everything is great.' Although the television camera had been shut down for lift-off, Aldrin filmed the operation, and only a few weeks later earth-bound spectators were able to experience the thrill of soaring up from the lunar surface for themselves.

As Eagle achieved lunar orbit at a speed of over 4,000 mph, Mike Collins was flying Columbia nearly 300 miles ahead of it. Columbia was not equipped to descend to the surface, but now, with Eagle in orbit, the first major hurdle of the return journey was over.

Two hours before, in the Sea of Crises, about 500 miles from Tranquility Base, the Russian mystery spacecraft, Luna 15 had either landed or crashed. Jodrell Bank reported: 'Braking signals on Luna 15 lasted for four minutes and then stopped suddenly. It definitely reached the surface of the moon, but it's not clear in what condition it has done so. It had a very high velocity on impact.' This news was not transmitted to Columbia or Eagle. The Russians, however, *were* publishing news of the American success, although they gave it relatively little coverage. Twenty minutes after Apollo 11 entered orbit, Tass reported the successful lift-off in a brief six-line bulletin.

The actual landing was reported quite fully in a broadcast on 21 July. It began, 'Comrades, we have just received a report that yesterday, at 23.18 Moscow time, the lunar module of the Apollo 11 spaceship landed on the moon in the Sea of Tranquility . . .' The Moscow Central Television broadcast of the first human steps on the moon spoke of the courage of the explorers. On the Moscow Home Service Academician A. A. Blagonravov commented on the American success with qualified admiration: 'All the scientific tasks involved in the present programme could, of course, have been solved to a considerable extent by sending automatic stations to the moon, and study of the magnetic field . . . could also . . . be solved successfully without the participation of man.' He emphasised the danger element of sending men to the moon, 'How big was the risk for the astronauts in the fulfilment of that programme? In every new undertaking there must be an element of risk, and the more complex the experiment, naturally the greater the risk. If a human being is used in the carrying out of any experiment the preservation of human life is, of course, very important and concern for this should be a matter of prime consideration. . . . I, like everyone else, wish those brave astronauts, whose courage one cannot but marvel at and admire, a successful end to their journey.'

Five hours after that broadcast Tass announced the Russian version of the fate of Luna 15: 'On 21 July 1969, the research programme in near lunar space and of checking the new systems of the automatic station Lunik 15 was completed. At 18.47 hours on 21 July a retro-rocket was switched on and the station left the orbit and reached the moon's surface in the pre-determined area.'

'Unlike the previous automatic stations Lunik 9 and Lunik 13, Lunik 15 could land in various areas of the moon's surface by changing the circumlunar orbit; two such changes of the orbit were made . . . and the new automatic navigation systems were tried out. During its flight in the orbit as a moon satellite, the station was used for scientific research in near lunar space and produced important data about the work of the systems on board the station, which are being processed.'

Whether that was all the Russians hoped their latest spacecraft would achieve is not clear: in the face of the American triumph it seems improbable that they would have admitted a total failure – but, considering the timing of their unmanned moon shot it seems equally improbable that they hadn't hoped to harvest bigger headlines with a more ambitious achievement.

Another Russian broadcast, by Vladimir Tregubov, emphasised Russia's alleged role in the Apollo moon conquest: 'It is quite appropriate that Armstrong and Aldrin should have left on our only satellite, among other objects brought from the earth to the moon, medals bearing the images of Gagarin, Komarov and Korolyev, those Soviet men whose great contribution to the conquest of space is incontestable, and without whose efforts and creative thought today's descent on the lunar surface by man, the representative of the planet earth, would have been simply impossible.'

The USSR may have been piqued at having to admit the Americans had won the race to the moon. But if they had to eat humble pie, so must the rest of the Communist bloc. Soon it was to be the turn of the USSR to chide another Communist country for not broadcasting the news: China! – even after splashdown the news of the lunar landing had apparently not been disclosed in China.

In the event of an emergency, Columbia could now descend to rescue Eagle's crew. On Columbia's 25th revolution of the moon, Aldrin reported he could see her flying high above Eagle. At 2.40 p.m. (EDT) the two craft disappeared behind the moon, and the vital rendezvous manoeuvres began. Then, after a virtually flawless mission, the first emergency cropped up. It is difficult to piece together precisely what happened for, even when communications were re-established with the spacecraft, the dialogue was so garbled that it was hard to hear exactly what was going on. Collins complained: 'Boy, you guys appear to be jerking around a bit.'

In the final stages of docking, with Eagle now ahead of him, Collins had had to steer Columbia's docking probe into the drogue on Eagle. It is designed to be guided into a socket at the bottom of this drogue, and three latches in the probe head lock the two modules together. Looking through the right-hand rendezvous window of Columbia, Collins lined up with the docking target aboard the LM. He then edged Columbia forward, but apparently did not sense the craft locking together – he therefore put Columbia into reverse to check if they had in fact docked. At that moment a thruster on Eagle fired for a fraction of a second. The probe hit the side of the drogue, and the two craft shook violently. Collins was a worried man: 'That was a funny one. You know I didn't feel us touch and I thought things were pretty steady. I went to retrack there, and that's when all hell broke loose.' Aldrin apologised: 'That thruster, it apparently wasn't committed.' Collins replied from Columbia, 'I was sure busy there for a couple of seconds' – in fact it was eight.

On earth, television commentators sensed that something must have gone wrong: up to that moment most of the mission had been carried out with split-second timing, but now the crucial docking was three minutes late. Then, at 5.40 p.m. (EDT) one clearly heard Mission Control referring to the spacecraft as Apollo. It meant that Columbia and Eagle had successfully docked.

The docking probe was now removed from the hatchway and the first two men on the moon crawled back to the comparative comfort of the CM. 'It's nice to have company,' commented Collins. Armstrong pointedly added, 'It's nice to have a place to sit down!'

Armstrong and Aldrin had thoroughly cleaned themselves and their equipment to guard against contamination before transferring the precious samples and other moon evidence to Columbia; they were about to settle down in preparation for the all important transearth injection burn of the service propulsion system. Four hours after docking they were scheduled to jettison the ascent stage of Eagle. Long before that, Aldrin reported hearing strange noises coming from the docking tunnel and the interior of the LM. Hastily the probe and drogue were stowed in the LM, and both hatches replaced ready for separation. Pressure had been built up inside Columbia so that any contaminated atmosphere would be transferred back into Eagle. The hatches were then sealed. A series of tiny explosive charges which surround the docking ring were fired, providing enough impetus to separate the modules, and Eagle floated slowly away into perpetual lunar orbit.

The rest of the mission was a welcome anti-climax back on

earth: to most people, the very fact that Apollo 10 had safely accomplished the docking manoeuvres around the moon meant it was a foregone conclusion that, from now on, Apollo 11 was plain sailing. Senator Edward Kennedy's car crash and Britain's possible entry into the Common Market commanded more space on front pages than the return of Apollo 11. The London *Times* devoted only one column to the alarming news of a 'Brief loss of contact with astronauts'. The all-important SPS burn (the firing of the service propulsion system's single rocket engine behind the moon, the astronauts' only hope of getting back to earth) received barely a mention. After a planned mid-course correction Houston found that Apollo 11, now in its 'barbecue mode', was rotating too fast. The computer had seemingly been wrongly programmed, and it was told to forget everything and begin again from scratch with a fresh set of instructions. Then ground control reported they had lost voice contact with Apollo. A mild panic broke out at Mission Control. 'Apollo 11, this is Houston broadcasting in the blind. If you can read me transmit on Omni Antenna' – an omnidirectional, low-gain antenna. There was no reply.

Fifteen minutes later Houston repeated the call. No reply. Then, after ten increasingly anxious minutes, contact was re-established. Apollo was returning safely.

Aboard the USS Hornet final preparations were being made to welcome the astronauts – a very different welcome from every previous mission: no hand-shakes, no red carpet – lunar contamination phobia had gripped NASA. Space officials, after considerable criticism for their 'dilatory attitude towards the potential perils of unknown moon germs erupting into cataclysmic epidemics on earth', had decided to play it safe. The first men on the moon would be greeted back on earth by a frogman administering a bucketful of disinfectant.

The eight-day mission ended on the morning of 24 July 1969 in the middle of the Pacific Ocean. Eight minutes before splashdown observers aboard the carrier Hornet saw a small red glow which quickly disappeared into the heavy clouds: Apollo 11 was cooling down after its base had been scorched white hot by the searing friction of re-entry into earth's atmosphere. The spacecraft had hit the narrow corridor which meant a safe return to earth. Dawn was breaking, and the carrier was riding in a heavy swell as the sound of the double sonic boom made by the descending spacecraft hit the ship. At 10,000 feet the main parachutes were released and the tiny capsule floated slowly down under its huge canopies of orange and white.

As if with a symbolic rebuke at Apollo's audacity, the ocean flipped the command module upside down, wrenching the astronauts over to hang beneath their couches from their safety harnesses. The official NASA description for this spacecraft position was 'Stable 2'.

President Nixon watched through binoculars as helicopters dropped naval frogmen to fasten a flotation collar around the capsule after it had been righted and was floating in position 'Stable 1'. The craft was sprayed down with a germ-killing fluid, and the hatch opened. Lieutenant Clancy Hatleberg handed the crew the bag of BIG's (a bag containing the Biological Isolation Garments which would seal them from the world until they entered the Mobile Quarantine Facility aboard Hornet), and all four men, bobbing about on the waves, started to scrub each other down. Lieutenant Clancy's breathing equipment filtered out any germs he might receive from the astronauts – whilst their masks filtered exhaled air to ensure earth's atmosphere remained pure.

As the opening of the hatch in an unprotected environment had already liberated some of the air they had been circulating for the whole return journey, this precaution seemed remarkably fatuous – but NASA presumably regarded the masks as a good public relations demonstration of health precautions.

The theatrical hygiene precautions were made even more absurd when the raft into which the astronauts had stepped from their capsule was sunk 18,000 feet to the bottom of the Pacific. If there had been any danger of contamination from moon germs, that raft might have been infected. In sinking it, NASA had potentially polluted the whole Pacific.

An hour after splashdown the astronauts had been lifted aboard the recovery helicopter and flown back to the safety of the deck of the carrier. They remained inside Helicopter No. 66 as it was towed to a lift and taken down to the hangar deck. Awaiting them was the mobile quarantine facility, a long shining metal caravan in which the astronauts were to be sealed to transport them, and any germs they might be carrying, to Houston, to the complex lunar receiving laboratory.

When the astronauts emerged from their helicopter there was no posing for photographs, no presentation of baseball hats – instead they quickly waved at the cameras and dashed into the quarantine facility. Then, two hours after splashdown, one of the most bizarre episodes of the Apollo saga began. Two marines marched to the end of the caravan and proudly stood to attention on either side of a small window inside which the curtains remained neatly drawn. They unfurled the Stars and Stripes, and the band began to play. President Nixon marched to the side of the window, and after an embarrassing pause, hands inside the caravan began to tug at the little curtain to pull it back. At last it was withdrawn, revealing the three astronauts stooping down to peer out at their President. Collins had grown a moustache. The comic-opera solemnity of the occasion was then fortunately relieved by the obviously sincere emotion displayed by the President as he excitedly welcomed the three men back to earth. 'I can't tell you all about the messages we received in Washington, but over a hundred foreign governments, emperors, presidents, prime ministers and kings have sent the most warmest messages we've ever received. They represent over 2,000,000,000 people on earth – all of them have had the opportunity through television to see what you have done.'

The President then began his world tour while the astronauts started a rather undignified journey to Houston, first by sea, then by air, carried inside their caravan aboard a Starlifter, and finally by road – towed slowly from Ellington Air Force Base to the lunar receiving laboratory in Houston. Even before the crew returned to Houston the precious rock samples were under close examination in the laboratory: they had been rushed back to Houston seconds after splashdown in case their condition deteriorated.

The primary examination of the samples was to attempt to categorise them before re-packaging, and despatch to 36 scientific groups around the world: these groups had been selected by NASA on the merit of their proposed experiments. To be as certain as humanly possible that the samples were neither contaminated nor allowed to pollute their surroundings, a special vacuum system had been built. It was the only one of its kind in the world: space gloves, operated by the sample examination team, reached into a vacuum chamber at pressures of 10^{-7} torr. (virtually as close to total vacuum as it is possible to go on earth).

In another section of the LRL germ-free mice were exposed to the lunar material and observed continuously for 21 days to note any abnormal changes in their health or behaviour. Periodically some of the mice were sacrificed for dissection to check any changes in their organs or tissue. In addition to these mice a number of other creatures were exposed to moon samples, including fish, birds, oysters, shrimps, cockroaches and houseflies.

Thirty-three species of plants and seedlings were exposed to study seed germination, and any abnormality in their health or the growth of plant cells. Special tests were conducted with six types of human and animal tissue cultures to isolate and identify any virus which might be present.

The crew themselves were also subjected to the most painstaking examination by a team of twelve doctors and assistants who lived with them inside the quarantine. All communication with the outside world was by telephone, and their first Press Conference was conducted behind enormous plate-glass windows.

Despite the fact that the mission had proved such a brilliant success there were several acrimonious exchanges between Houston and the spacecraft during the transearth coast about the lack of documentation on the samples. Even though Armstrong and Aldrin had overstayed their planned time on the moon (causing angry shouts, and later almost plaintive pleas from the Capsule Communicator as he tried to persuade them to return to Eagle), they had not collected as much lunar material as intended. At first it was thought they had only gathered 20 lb.: in fact they returned with 70 lb., but much of it had been hurriedly grabbed in the last few seconds of EVA, and no one knew exactly where it had come from. Armstrong had reported that the descent engine exhaust had made almost no crater, but had scorched a star-shaped burn of radiating lines directly below the LM. He had collected his bulk samples from between these lines and told Mission Control he suspected the material to be only minimally contaminated by Eagle's rocket engine. The degree of contamination will only probably receive a final evaluation at Christmas 1969: after preliminary examination, samples were sent out from the lunar receiving laboratories in Houston to the many experts around the world who were given ninety days to report their findings. Because Apollo 11 had returned with less lunar material than had been hoped, samples were rationed even more strictly than originally planned. Foreign research teams had precedence for diplomatic reasons – much to the anger of some American scientists who had spent thousands of dollars building facilities for analysing their quota of the moon. Even before the official analysis was published some facts emerged after the first spate of 'revelations' about the 'glass moon'.

The moon seemed to be of basaltic origin (as several experts had prophesied), with a crust about twelve miles thick, and a molten core. Apart from its almost total lack of atmosphere, it was geologically remarkably similar to earth. Volcanic action had probably reduced impacting meteorites into basic elements, and the surface was covered with igneous material, including layers of glass particles. As Armstrong had noted, many samples resembled fragments of pumice, with tiny holes in the rock where heat had expanded gases into vitreous bubbles. The precious metal titanium was found in much greater quantity than on earth – but probably in quantities which will still make it uneconomic to fly back to earth. Sadly, there were no reports of any cheese.

Seven years ago Dr von Braun was asked why go to the moon? He replied, 'We go because the moon is there. It is a great challenge and we have means of meeting the challenge. Most of man's scientific accomplishments have resulted from curiosity with no specific goal foreseen. We know that man's spirit must now, as always, accept the challenge of the unknown; and from past experience we know that it will be rewarding beyond our present ability to imagine. The Apollo moon project is not the end.' Just before the Apollo moon-shot I asked him, therefore, what he considered lay beyond the moon landing. He answered, 'I can speak only from the standpoint of an engineer about what is feasible and what would appear to be logical options for the country to pick up. The decision ultimately whether the country feels it should support such a programme is up to the tax-payer, Congress and the President.'

'It seems pretty obvious that if we are successful to land an American on the moon with Apollo 11, and bring the fellows home alive, that we have done this programme with less hardware than we originally thought was necessary. As a result we will have a sizeable fleet of vehicles, capable of carrying more people to the moon, in the warehouse ready to go.'

'Our present plan therefore is nine more Saturn V Apollo flights to the moon after the first successful landing. In addition we have planned to build a rudimentary space station in earth orbit for a variety of manned activities. That will include, for instance, meteorology, astronomical research from an observatory in orbit, biological research, and something we find particularly exciting, the gradual development of an earth resources programme.' That programme is designed to study the earth from space, evaluating where food is grown, and how successfully, how cities are growing, and therefore how many mouths must be fed. It has given rise to a new science and could have come in the nick of time with the world facing both a population explosion and growing food shortages.

Von Braun is very conscious of the charges of costly extravagance that have been levelled at the space programme: 'Of course, we must make space flight cheaper. It's very important to reduce the present cost of about 500 dollars per lb. payload in earth orbit to something like 50 or even 30 dollars. Our next objective in the transportation end of space flight will be the creation of a re-usable vehicle that can fly into an earth orbit and back – many times out and back just like an airplane. For such a vehicle a drastic reduction of costs by a factor of ten is entirely feasible.'

As von Braun implies, the immediate future of space flight, and particularly of Apollo, has become a matter of economic viability. Despite that, I was surprised, perhaps unjustifiably considering the consistently high standard of the periodical, to read an editorial in *The Economist* entitled **Frontier Moon** which summed up the achievement of Apollo 11 better than any other comment I have read: the writer described space as unexpectedly friendly: 'Maybe it is short on such usually considered essentials of life as breathable oxygen, but it has produced no snakes yet, no dragons, no yellow fever, no elemental storms that claimed the lives of so many terrestrial explorers. It is turning out to be nothing like as bad as we thought.'

'Evolution is normally a gradual process in which the apes grow imperceptibly less ape-like and more man-like with succeeding generations. But once in aeons the skin of history gives a twitch and the processes of millions of years are condensed into a few minutes. Man is doing that now, venturing out of his normal environment to step on to a new planet. The last time such an evolutionary jump occurred was when the first fish ventured on land and found that the air was sweet.'

Above *Roll-out of 'The Big One', May 1969.*

Top left *Edwin Aldrin* (*Lunar Module Pilot*). Top right *Neil Armstrong* (*the first man on the moon, and the Apollo 11 Mission Commander*), and right *Michael Collins* (*the Command Module Pilot*).

Apollo 11
1.9 secs after
ignition

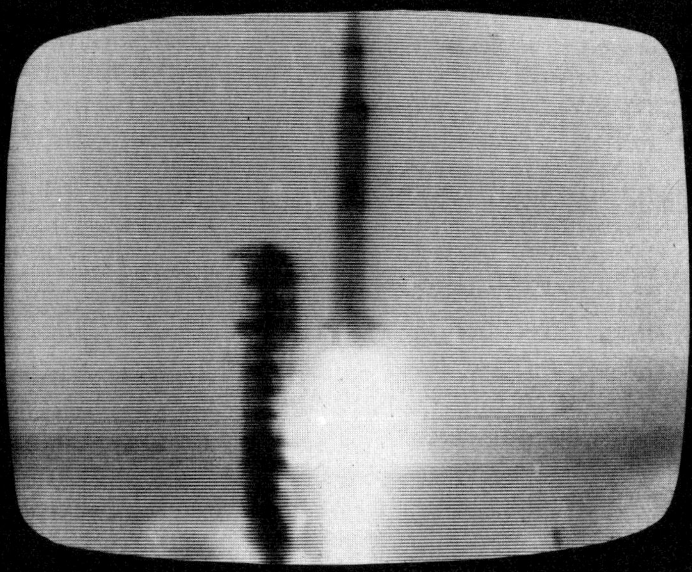

Lift-off to
the Moon

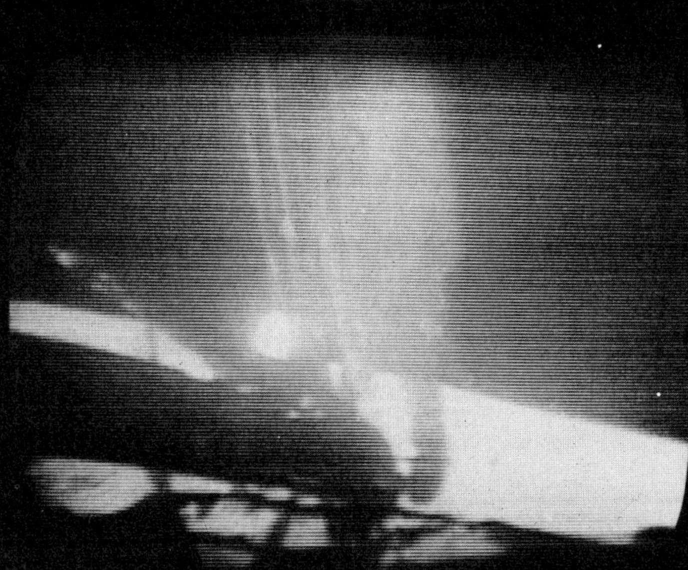

'... one giant
leap for
Mankind'

Magnificent Desolation'

Aldrin moonwalks beside the Stars and Stripes

Neil and Buzz, I am talking to you by telephone from the Oval Room at the White House'

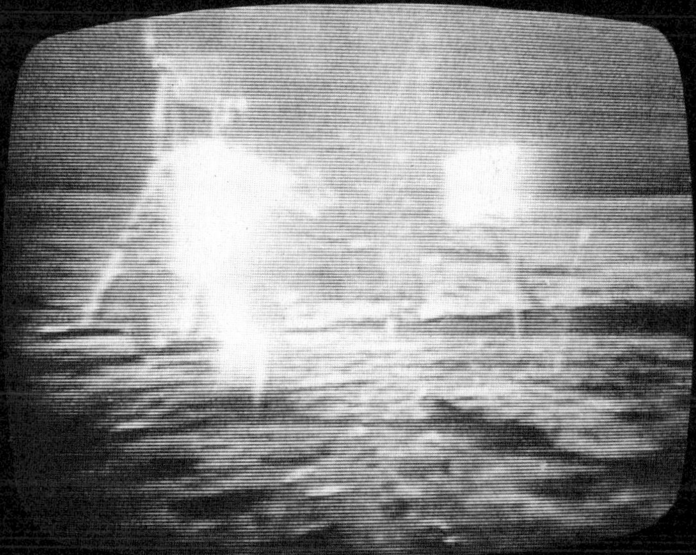

FLIGHT PLAN

CSM		LM		MCC-H
CMP		CDR	LMP	
	0130 EDT 112:00	PREP FOR CABIN DEPRESS CONNECT OPS O_2 HOSES LMP DON HELMET CDR DON HELMET CONNECT PLSS H_2O HOSES DON GLOVES		
EAT PERIOD (1 HOUR)	112:08			DUMP DSE
CREW STATUS REPORT (SLEEP) SELECT COMM NORMAL LUNAR CONFIGURATION	MSFN	PRESSURE INTEGRITY CHECK PLSS O_2 ON	SET CHRONOMETER	START EVA 0+00
	112:30	FINAL PRE-EVA OPERATIONS DEPRESS CABIN FINAL SYSTEMS CHECKS	OPEN FWD HATCH PLSS H_2O ON	
	112:39	INITIAL EVA EGRESS TO PLATFORM RELEASE MESA DESCEND LADDER REST/CHECK EMU SYSTEM	ASSIST AND MONITOR CDR TURN TV ON ACT 16mm CAMERA	0+10 UPDATE CSM LM ACQUISITION TIME
COPY PAD (SEE GET 114:20)	112:45			
		ENVIRONMENTAL FAMILIARI-ZATION CHECK STABIL, MOBIL, EMU	MONITOR CDR OPERATE 16mm CAMERA	0+20
	113:00	CONT SAMPLE COLLECTION COLLECT AND STOW SAMPLE		0+30

MISSION	EDITION	DATE	TIME	DAY/REV	PAGE
APOLLO 11	FINAL	JULY 1, 1969	112:00 - 113:00	5/19	3-79

FLIGHT PLANNING BRANCH

FLIGHT PROFILE

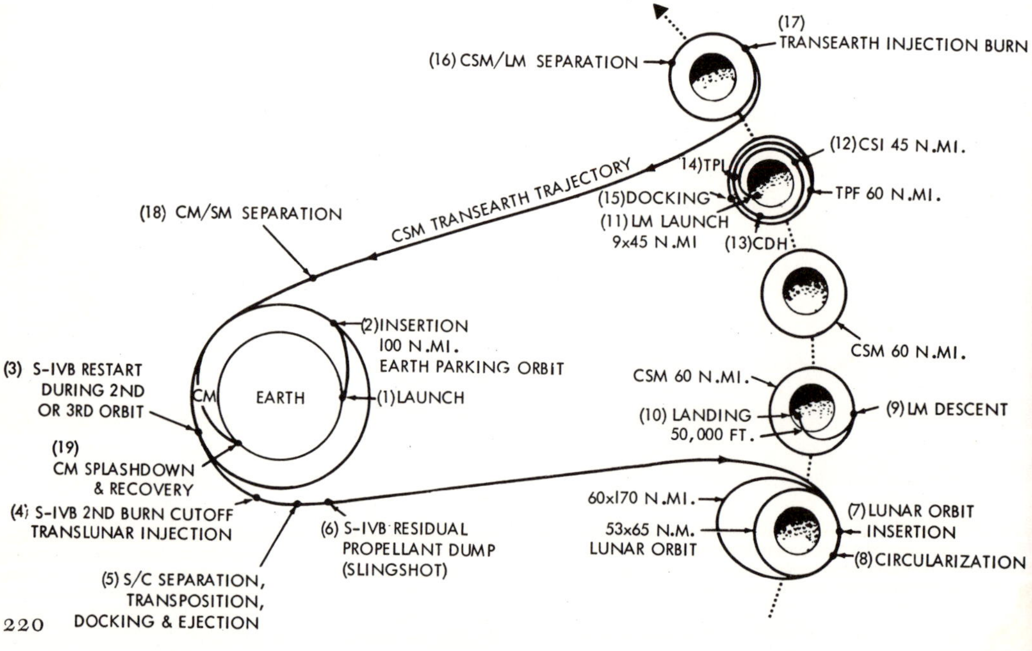

(17) TRANSEARTH INJECTION BURN
(16) CSM/LM SEPARATION
(12) CSI 45 N.MI.
(14) TPI
TPF 60 N.MI.
(15) DOCKING
(11) LM LAUNCH 9x45 N.MI
(13) CDH
CSM TRANSEARTH TRAJECTORY
(18) CM/SM SEPARATION
CSM 60 N.MI.
(2) INSERTION 100 N.MI. EARTH PARKING ORBIT
(3) S-IVB RESTART DURING 2ND OR 3RD ORBIT
CM
EARTH
(1) LAUNCH
CSM 60 N.MI.
(10) LANDING 50,000 FT.
(9) LM DESCENT
(19) CM SPLASHDOWN & RECOVERY
(4) S-IVB 2ND BURN CUTOFF TRANSLUNAR INJECTION
(6) S-IVB RESIDUAL PROPELLANT DUMP (SLINGSHOT)
60x170 N.MI.
53x65 N.M. LUNAR ORBIT
(7) LUNAR ORBIT INSERTION
(8) CIRCULARIZATION
(5) S/C SEPARATION, TRANSPOSITION, DOCKING & EJECTION

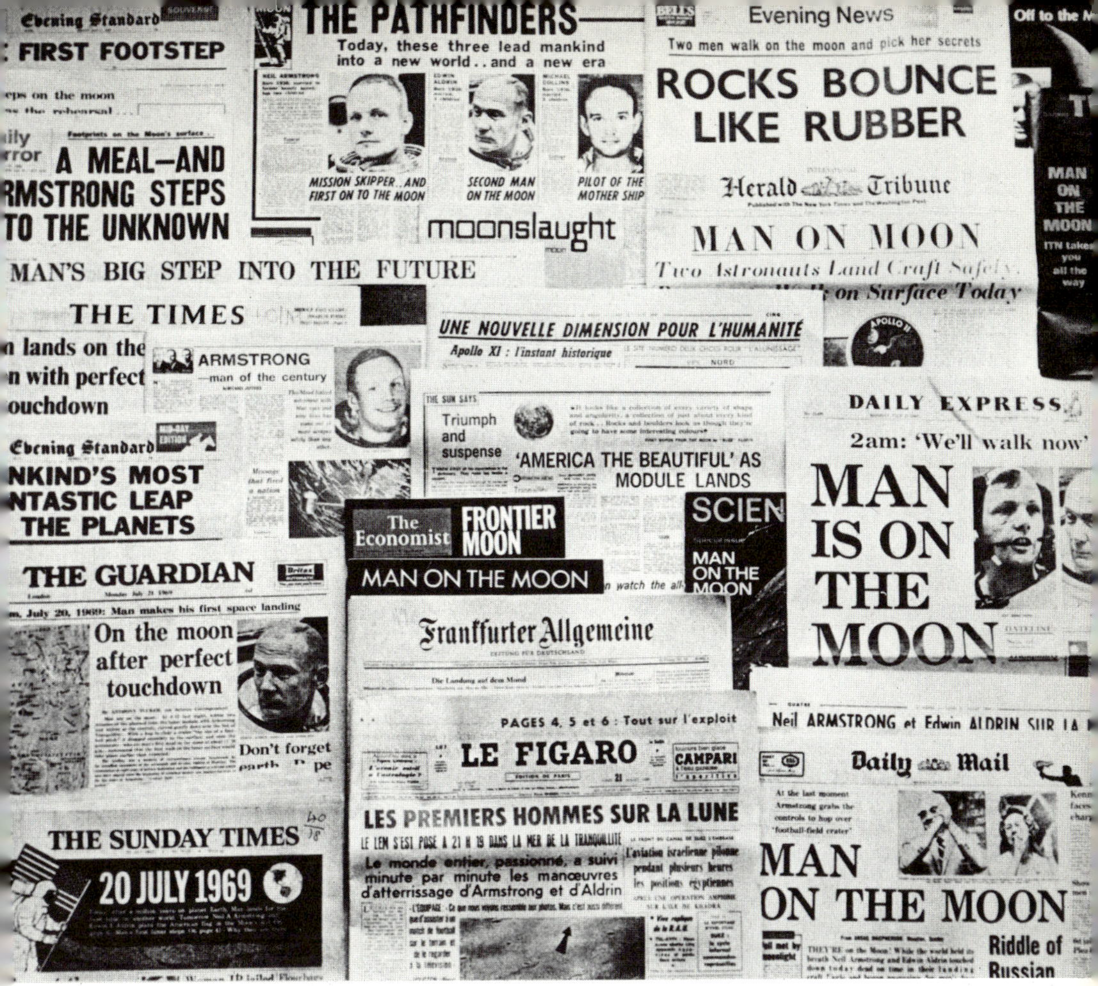
Evening Standard
FIRST FOOTSTEP
A MEAL—AND ARMSTRONG STEPS TO THE UNKNOWN
MAN'S BIG STEP INTO THE FUTURE
THE PATHFINDERS—
Today, these three lead mankind into a new world . . and a new era
MISSION SKIPPER...AND FIRST ON TO THE MOON
SECOND MAN ON THE MOON
PILOT OF THE MOTHER SHIP
moonslaught
Evening News
Two men walk on the moon and pick her secrets
ROCKS BOUNCE LIKE RUBBER
Herald Tribune
MAN ON MOON
Two Astronauts Land Craft Safely
Off to the M
MAN ON THE MOON
ITN takes you all the way
THE TIMES
UNE NOUVELLE DIMENSION POUR L'HUMANITÉ
Apollo XI : l'instant historique
ARMSTRONG
man of the century
THE SUN SAYS
Triumph and suspense
'AMERICA THE BEAUTIFUL' AS MODULE LANDS
DAILY EXPRESS
2am: 'We'll walk now'
MAN IS ON THE MOON
Evening Standard
NKIND'S MOST NTASTIC LEAP THE PLANETS
The Economist
FRONTIER MOON
MAN ON THE MOON
SCIEN
MAN ON THE MOON
THE GUARDIAN
July 20, 1969: Man makes his first space landing
On the moon after perfect touchdown
Don't forget
Frankfurter Allgemeine
PAGES 4, 5 et 6 : Tout sur l'exploit
LE FIGARO
CAMPARI
LES PREMIERS HOMMES SUR LA LUNE
Le monde entier, passionné, a suivi minute par minute les manœuvres d'atterrissage d'Armstrong et d'Aldrin
Neil ARMSTRONG et Edwin ALDRIN SUR LA
Daily Mail
MAN ON THE MOON
Riddle of Russian
THE SUNDAY TIMES
20 JULY 1969

APOLLO 11

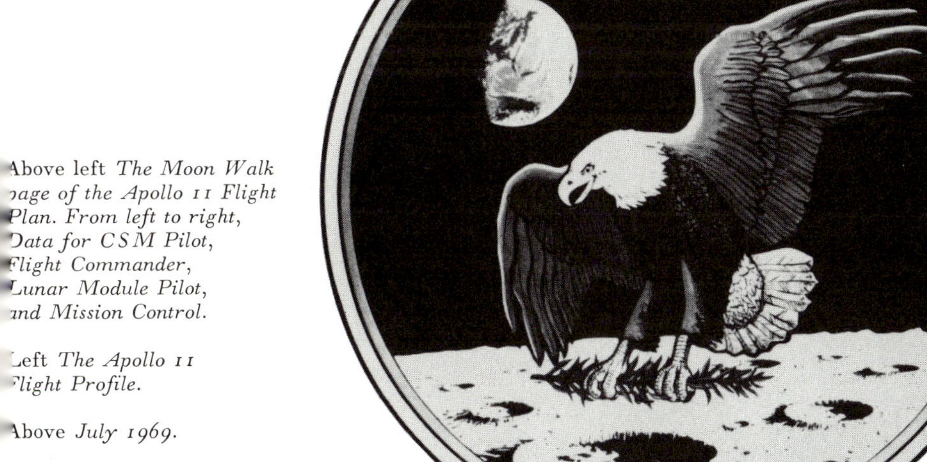

Above left *The Moon Walk page of the Apollo 11 Flight Plan. From left to right, Data for CSM Pilot, Flight Commander, Lunar Module Pilot, and Mission Control.*

Left *The Apollo 11 Flight Profile.*

Above *July 1969.*

Right *The Apollo 11 insignia. The American Eagle plants an olive branch on the Moon.*

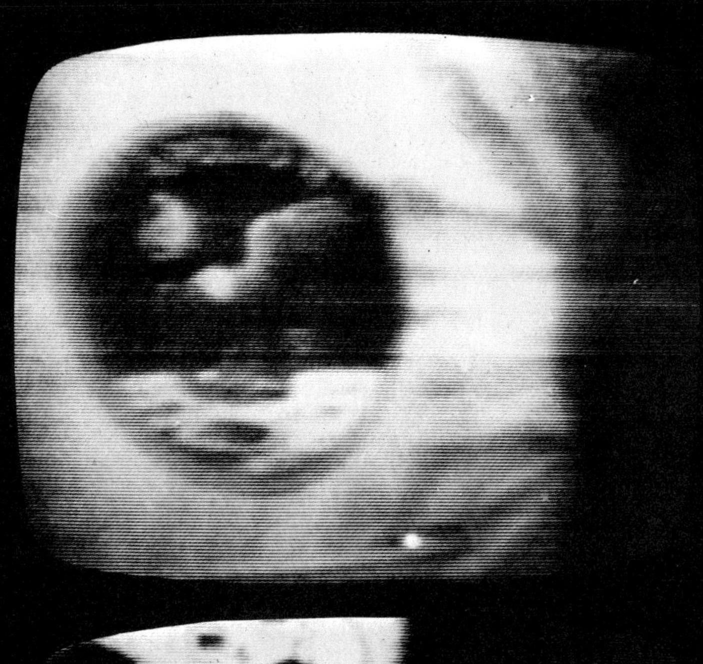

The Apollo 11
Insignia from
Space

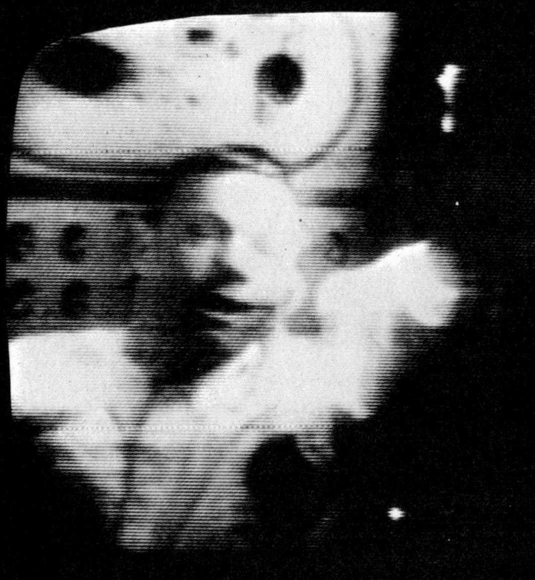

The man who
never saw the
lunar landing.
Mike Collins
Command Module
Pilot

The First Man
on the Moon.
Neil Armstrong
Apollo 11
Flight Commander

he Second Man
n the Moon.
uzz Aldrin
unar Module
ilot

ission Control
ouston after
plashdown

table One' -
t last, after
ight minutes of
table Two' -
pside down!

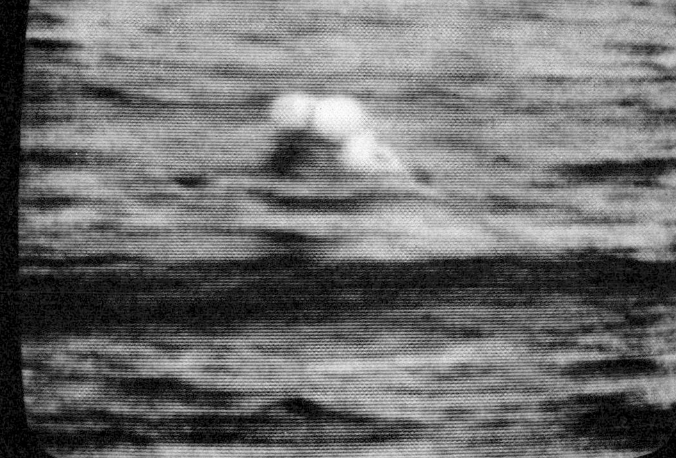

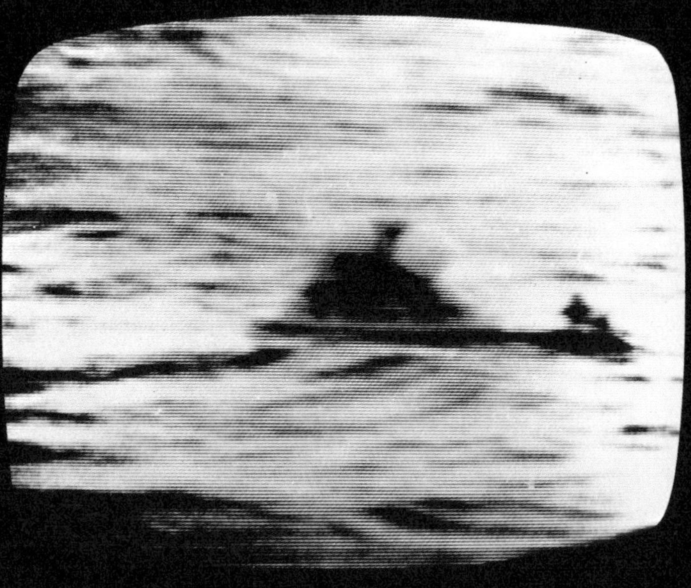

Three men in a boat, plus the Big Swimmer, Lieutenant Hatleberg

The Man on the Moon is winched into quarantine

66 returns its third Apollo crew to Hornet

66 is
rolled to the
Mobile Quarantine
Facility

he Moon Men
n their BIGs
ash to the
uarantine
aravan

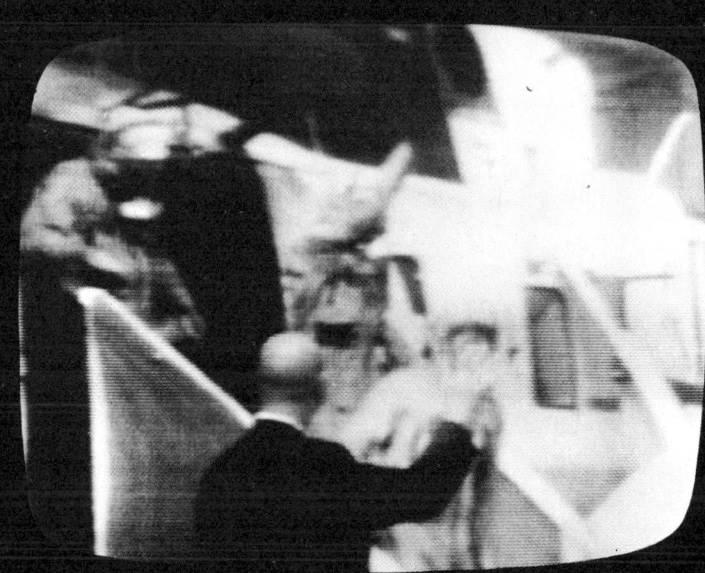

eft to Right:
rmstrong, Collins,
ldrin and
resident Nixon as
e welcomed them
ack to Earth,
hrough the window
f the Quarantine
acility

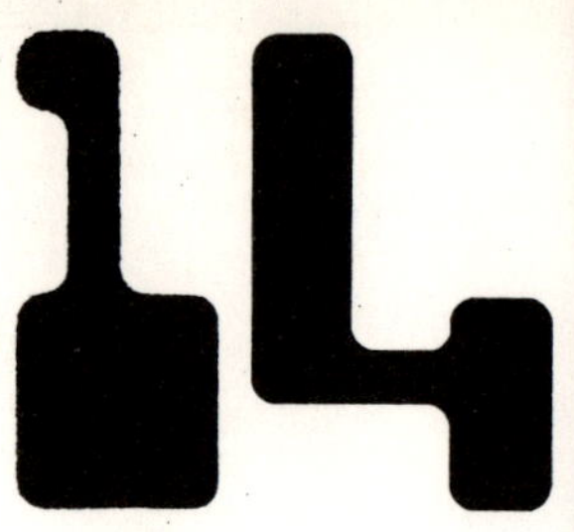

The Moon and Beyond The value of man's greatest technological adventure and the future challenge of interplanetary and intergalactic travel

'When history looks back on this period, it will be obvious that it was a good thing there was a race. We need space technology to solve problems we have here on earth.' The scientist and science-fiction writer Arthur C. Clarke said those words in 1968. Today there are still many scientists who disagree with him about the merits of both space technology and the space race. The 'Man to the Moon' prestige struggle has also been heavily criticised. Sir Bernard Lovell's comments before Apollo 8 typify the attitude of many serious scientists and astronomers to manned expeditions.

Both the Apollo and Soyuz missions have been attacked on the grounds that they kept a disproportionate number of scientists occupied on a project of allegedly secondary importance: in the US six per cent of the scientific community is involved with the space programme. The exploration of space is constantly abused as being little more than a colossally expensive exercise in vulgar public relations; many see it as a naïve bid for international prestige rather than as a worthwhile expansion of the frontiers of science. Certainly even the most ardent supporters of America's space effort must admit that President Kennedy stipulated prestige as one of the main reasons for committing the United States to put a man on the moon 'before the end of this decade'. However, perhaps the strongest criticism of expenditure on space flight comes from those who feel that the money and effort would have been better invested in solving the social and technological problems of the earth before looking elsewhere. Why look for trouble when one is surrounded by it? The Archbishop of York, Dr Donald Coggan, urged cuts in the vast sums spent on space, asking is it 'the best use of resources at this point of human history when ignorance and disease stalk the earth?'

Today the population of the world is increasing at the rate of 200,000 every day, yet 200,000,000 people are already living at starvation level. Ironically the population explosion is not due to an accelerating birth rate, but to modern medicine. By the year 2000 the world's population will have more than doubled. It is against facts like these that one must judge the wisdom of 'man's greatest adventure'. It is tempting to be dogmatic and say that space exploration is either wholly futile and a waste of money, or as useful to humanity as Columbus' voyage of discovery and a challenge we cannot ignore. The truth is rarely a compromise, but it may not be until the year 2000 that one will know for sure whether the right decisions were made half a century before. The nature of man himself must be taken into account, and as Professor Dennis Gabor pointed out in his excellent book, *Inventing the Future*, the self-annihilation of mankind itself may be avoided by diverting high-power brains from armaments research into space research.

The total cost of the American Apollo programme has so far been about $24,000,000,000; by way of comparison Mercury cost a mere $500,000,000, and Gemini $1,500,000,000. It is impossible to obtain truly accurate comparative costings for the Russian programme, partly because of the difference between Communist and Capitalist economic structures. The Russian space industry reveals neither its final objectives nor its balance sheets. However, a consensus of intelligent guesses would put their bill for making a manned lunar landing possible at about $15,000,000,000. To get such astronomical sums of money into perspective, the NASA budget for 1969 was $3,854,000,000; the Vietnam War effort was allocated about eight times that sum. Altogether the US has probably spent about $50,000,000,000 on space, but as the economist J. K. Galbraith often says, 'If you want more money, you can always print it.' The real expenditure on space can only be measured in terms of the human and material resources devoted to the task. At its peak the Apollo programme was employing full-time 400,000 men and women in factories and laboratories, and another 10,000 in universities. Millions of dollars worth of rare metals and fuels etc. have literally vanished into thin air. Whether or not either the manpower or the materials would have been put to better use if they had not been needed for Apollo is still a hotly debated point: the historical fact is that the Apollo project created many work opportunities in previously depressed areas, and pioneered significant scientific achievements at the same time.

To answer questions of cost-efficiency the jargon makers of NASA coined the term 'spin-off'; even before the flights to the moon, this was considerable. The most quoted examples of technological 'spin-off' are non-stick frying pans and Velcro fasteners, but such trivia tend to distract from the major advances in fields like micro-electronics and computers, where research has been enormously accelerated by the space race: hardly any aspect of science has not benefited. It is invidious to select examples of hardware for praise, as they will no doubt continue to stem from the pure research work which has already been completed. Medical engineering owes a great debt to space spin-off: new alloys and bearings have greatly improved the efficiency of prosthetic limbs: tiny implants for measuring blood-pressure have been devised small enough to be injected into arteries and carried to the heart itself; a meteorite impact sensor has been adapted to detect the tiniest muscle tremors, and thereby help an early diagnosis of maladies like Parkinson's disease. A whole range of bio-medical devices intended for monitoring the health of astronauts, has now become available to the medical profession. Techniques of achieving ultra-sterility have been devised in attempts to avoid the contamination of outer space, and there have been many advances in cryogenics (the science of very low temperatures) for the handling of rocket fuels.

Absolute dependability is essential in a rocket, and for this reason the study of fluidics grew: fluids pass through complex circuitry and switching mechanisms in a similar manner to electricity, but with the advantage that fluid circuitry is more rugged and less prone to damage than electronic valves, transistors, and even solid circuitry.

The electromagnetic tools developed to smooth the weld seams on Saturn V are now in use in the shipbuilding and automobile industries. The design of the miniature television cameras which show the separation of rocket stages caused breakthroughs in design, as did the method of 'tidying' the pictures transmitted by Ranger and Surveyor, using a computer. NASA claims that more than 2,500 similar 'hardware' innovations have been the by-product of space technology.

The exchange and cross-fertilisation of ideas between many branches of science has been another considerable benefit stemming from the space race: because of the intricacy of many of the problems of survival in space, the traditional barriers of technological specialisation have had to be broken down. For the first time for many years biologists, electronics engineers, doctors, physicists, neurologists and mechanics have been forced to communicate with one another, causing an interchange of scientific ideas which has rarely been possible since the Renaissance, when men like Leonardo da Vinci knew no specialist frontiers, and worked simultaneously on aeronautics, astronomy, anatomy, and engineering.

It is perhaps unfair to count the information gained from all the orbiting satellites as a credit to the space-flight balance sheet, for they would probably have been launched irrespective of any manned or robot space exploration policy: however they have added considerably to our knowledge of geophysics and our abilities in weather forecasting, navigation, and communication. If they ever make it possible to forecast the weather infallibly for just five days ahead, the annual savings to the United States alone will be gigantic: $3,000,000,000 in water supplies, $2,500,000,000 in agriculture, $100,000,000 in transport, $75,000,000 in the retail trade, and $45,000,000 in the timber industry. Apart from the manned earth-orbit station, at the time of writing no major objectives have been finalised by the Americans to follow the moon landings; the Russians, with typical caution, state vaguely that they aim to expand man's scientific knowledge of the universe. Both sides however are intent on analysing more lunar rock to decide once and for all where the moon came from – whether it was once part of earth, a meteoroid, or formed in some other way. Theoreticians still argue, but such a discovery will eventually lead to a much greater understanding of the origin of the earth itself, the sun and our solar system. The lack of atmosphere makes the moon an ideal site for optical telescopes, and the low gravity and absence of wind mean that huge radio telescopes can be built. Astronomical studies from the moon should make the discovery of the origin of the universe only a matter of time. Studies of the sun and the way it liberates energy could bring the elusive dream of nuclear physicists a stage closer to reality – the control of nuclear fission to generate electricity directly. Already the discovery of the Van Allen belts by Explorer I, and the earth's magnetic 'tail' by Lunik 10, have greatly contributed to understanding how energy can be contained and transmitted.

Deep drilling could disclose the moon as being an Eldorado of mineral wealth: the earth is still revealing secrets to geologists, so it may be many years before a definitive review of the moon's resources can be published. The first findings from Luniks 9, 13, and Surveyors 5, 6 and 7 were undramatic – they showed that it was composed of materials similar to basaltic rock, as found on the ocean bed – but scientists hope that it will reward them with surprising secrets for many years to come; it is those secrets that provide the excitement, and could also bring scientific bonuses which, even today, it is still impossible to predict or assess. An eminent consultant to NASA on minerals, Professor George Mueller, was confident enough to prophesy a year before the moon landing some of the advantages of going there: 'According to my predictions, the moon will be essentially extremely rich in mineral deposits which are associated with basaltic rocks, which means copper, mercury, silver, and arsenic; and rather deficient in those metals which are associated with granitic rocks – namely uranium and tin for example. Another source of minerals on the moon will be impact minerals originating from meteorites. Iron meteorites are very valuable objects from the mining point of view. They contain 90% iron, about 9% nickel and 1% cobalt – also they are maybe 10- or 100-fold more rich in platinum metals than the richest known terrestrial platinum deposit, and can contain 10,000-fold higher concentrations of diamonds than the best deposit known on earth.

'We have very good reason to believe that they would be in a much higher proportion on the surface of the moon than on the surface of the earth. The earth's atmosphere has a tendency to corrode meteorites and disperse them all over the surface once they fall, but they reach the moon in their entirety – in a much more complete form.

'Once we have invented the atomic rocket I do feel that lunar exploitation will very soon become an economic possibility, particularly in the case of iron meteorites which have impacted on the moon and might be of very considerable size; there is a possibility, I believe, that an atomic rocket could have a device which just picks up the whole meteorite and brings it back from the moon to the earth. In this way the cost of mining would be drastically reduced.

'We are running out of high-grade mineral deposits within the earth's crust and we are being forced to utilise lower and lower grade minerals, which are of course more costly to exploit and will cause a rise in the prices of all our metals. I believe for this reason we will be forced to resort to lunar and other extra-terrestrial mineral deposits in the future.'

Some space experts would argue against the moon ever being exploited because of the high cost of transporting the mineral deposits back to earth, however small and valuable they might be. Nevertheless, the Soviet scientist Dr Ari Shchernfeld believes that it may be possible to build a kind of 'space lift' to ferry cosmonauts and goods to and fro. He envisages a spaceliner flying an orbit which loops both the earth and the moon: its trajectory would be planned to come close to the moon from time to time by using correcting rockets. Cosmonauts and materials would travel up to the spaceliner and rendezvous using a smaller rocket to carry them. 'While flights to the planets could be made only during interplanetary navigation seasons, flights to the moon could be made at any time – weekly or even daily. It is possible to imagine a huge spaceliner with all the comforts that are now provided by ocean liners. Rockets with loads for building such a "space-lift" could be launched from a space station orbiting the earth, a prototype of which was already created when Soyuz 4 and Soyuz 5 were linked up in space. The space station used to build a lift would also be a huge composite satellite created by linking up space vehicles launched from earth.' (*Novosti*, 23 January 1969).

Before such moon traffic can begin, however, there is considerable wealth orbiting the earth which may be worth retrieving – and equally, might need cleaning up before Dr Shchernfeld's daily moon excursions can begin. Today, in 1969, there are already over 4,000 spacecraft and other objects adrift in space or in long-term earth orbit; the majority of them no longer perform any useful function. At the 19th Congress of the International Astronautical Federation in October 1968 it was estimated that by 1980 about $5,000,000,000 worth of junk will be in orbit. By then it may become economically worthwhile for a cosmic 'rag and bone' man to collect it all together and avoid the possibility of a collision with a future manned spacecraft. Statistically speaking such collisions are almost impossible, but the first one (fortunately between unmanned vehicles) has already happened: in March 1965 two communication satellites launched simultaneously by the United States Navy drifted together two months later, and their aerials clashed together 'like crossing swords'. No damage was done, but as inner space gets more crowded the possibility of a very expensive collision will grow.

Not only inner space, but the moon itself is now becoming subject to sophisticated litter. So far 18 spacecraft have landed – probably ultra-clean, but junk nonetheless, for it is very unlikely that such robots could be in any way incorporated in future lunar bases.

Both the USSR and the United States have already carried out research on prototype lunar bases in the Antarctic. However the lack of water and air, plus the weight limitations on lunar expeditions will make the moon very much more difficult than even Antarctica to make habitable. Using today's Saturn V as a freight carrier the weight limit is 25,000 lb. and even if the rumours about Russia's 10,000,000-lb. thrust rocket prove correct, it will only exceed the American freight payload by 10,000 lb.

Until it is known with certainty that useful elements can be extracted from moon rock, the first lunar base will have to be entirely self-contained. It will probably consist of 25,000-lb. modules, each performing its own specific function. A study commissioned by NASA proposes the following combinations: a module for crew quarters; a life-support module containing oxygen supplies, air-conditioning, and water/waste recirculation; a power module containing a nuclear reactor to provide electricity; a communications module with a transmitter and a computer unit; a fuel module with fuel for moon-based vehicles; a maintenance module for equipment repairs, and a shelter module into which personnel could retire during unusually dangerous meteorite showers or solar flares.

All these modules would be connected by tunnels through airlocks so that a meteorite penetration of one of them would not depressurise the entire system. Eventually such modules would be covered with lunar soil so that astronauts could immediately have somewhere safe to live while they assembled the other sections of the system. Such modules are expected to cost about $125,000,000 each to carry to the moon in the early 1970's, but by 1980 the cost may have dropped to $50,000,000. Assuming chemistry keeps pace with the other branches of space science, the requirements of a manned base, apart from food, can largely be met using carbon, oxygen and hydrogen: combinations of these form air, water, and rocket fuels. All three are almost certainly abundant on the moon, but unfortunately in very stable forms which require considerable energy to liberate from the lunar rock: it is still not known whether such an operation will be practical. One experiment on extracting water from rock in a lunar simulator using solar heat was encouraging, but large quantities of rock had to be crushed before the extraction could begin, and such an operation would necessitate an enormous power supply. Some scientists still believe that water exists on the moon in the form of ice in underground caverns. If they are correct, water supply will be an easy matter. If they are not, and worse, if future missions prove the moon to be basically of meteoric origin, the difficulties of establishing bases may be almost insuperable, for meteoric rock contains only 0·1% water: volcanic rocks on the other hand can contain as much as 5% water, once processed.

With no atmosphere, an estimated 200 tons of meteorite material falling on its surface every 24 hours, temperature variations between +250 deg. and −250 deg. F., and the occasional scorching solar flare, the moon, to say the least, is not ideally suited for human habitation. However, even in 1963 NASA had faced the problem seriously enough to invite all the major aerospace contracting companies in the USA to submit general ideas for solutions.

The simplest answer to the temperature extremes was a proposal for a base 100 feet underground, where the temperature is thought to remain at a steady −10 deg. F. An alternative was finding a site near one of the lunar poles where the surface temperature is more constant. Many suggestions for surface installations made use of the variable heat absorption of different surface textures: for example a spherical building was proposed which would present a highly polished reflective face to the sun at lunar 'noon', and a coarse black face to the sun at lunar dawn and dusk. One company is even developing a chameleon paint which automatically changes colour as the heat of the sun increases, absorbing a constant amount of warmth.

For radiation protection, use will be made of the physical property of similar electrostatic charges, which repel each other. The dangerous radiation particles which form most primary cosmic radiation are positively charged nuclei. It is hoped that by positively charging the surface of a lunar base it will be possible to repel these particles, or at least slow them down sufficiently to render them harmless.

Meteorite damage may be more difficult to avoid. Both Russia and America have already developed protective devices for spacecraft, and these may be applied to the lunar base roofs. Inflatable shielding has been suggested, but many experts consider such structures too vulnerable. The Russians have partially avoided the problem with a detailed proposal for a three-tier underground base complete with laboratories and a nuclear power station.

Much of the food will be grown hydroponically, whereby plants, suspended in water, live on an artificial chemical diet. During the night it is thought that, using reflectors, 'earth-light' will be sufficient to keep them growing – the earth is many times brighter than the moon seen from earth.

Initially manpower on the moon will be absurdly expensive – $80,000 per man-hour, reducing to $50,000 by 1980. Because of this most of the structures for the moon will be prefabricated on earth, but before complex lunar living quarters are established, transport devices will have to be taken to the moon to make the first exploratory journeys possible.

Even at lunar gravity walking in a space-suit can be exhausting: after the geological surveys in the immediate vicinity of the spacecraft, the first explorers will venture as far as the visual and radio horizon – a distance of about 5 miles. They will probably travel in a kind of two-man lunar taxi. Since 1966 a large variety of lunar transport vehicles has been designed for consideration by NASA, all trying to overcome the problem of not really knowing much about the surface on which they have to travel, but assuming it to be rough terrain. One looks like a primitive piece of agricultural machinery, another like a bunch of walking balloons. Later astronauts will travel up to 25 miles from the landing site, using devices whose prototypes have already propelled James Bond into some improbable situations. A two-man version weighing 400 lb., powered by 5 rocket engines will fly across the moon at about 20 mph. The most advanced design so far for lunar surface travel is the Mobile Laboratory, or Molab for short: it is designed for a 14-day journey, carrying two astronauts and a considerable amount of scientific equipment. It is too heavy to fly to the moon with its occupants, and therefore is designed to be landed in advance on a modified version of the lunar module, called the LM truck. The LM truck has extendable ramps for off-loading the Molab; on arrival the Molab unfolds and automatically rolls down to the surface to await its passengers. As the Molab has a pressurised cabin the astronauts, once inside, are able to work without space-suits. Out of radio contact with their own LM, they communicate directly with earth. Complex tasks of rock sampling, photography and mapping have already been detailed for Molab, and a rocket chair completes its equipment, enabling a space-suited astronaut to hurry back to the LM across any obstacle in case of emergency. The maximum range of the Molab is about 300 miles; it benefited considerably from the design studies for a cancelled programme called Prospector, an early project for an unmanned Molab.

Because of the extreme temperature changes most designs for lunar surface vehicles have dispensed with pneumatic tyres, and overcome the shock absorbing problem with enormous wheels held to the vehicle by giant springs which spiral from the hub to the wheel rim. One such six-wheeled juggernaut is capable of surmounting a 10-foot wall and crossing a 10-foot chasm, each wheel being driven independently by a fuel-cell powered electric motor.

Despite the much publicised campaigns by the US and the USSR for the peaceful use of space, there is still a chance that the first moon base could have a military purpose, although the degree of 'overkill' already possessed by both sides with their ICBM's could reduce that possibility. Both countries are signatories to the UN Treaty of 27 January 1967 which lays down that activities in space must be for the benefit of all nations, and that no claim of sovereignty over an extra-terrestrial body will be entertained; a different agreement specifically excludes nuclear or other mass destruction weapons from space.

Already there are signs that the United States could easily return to the Russian practice of linking space-flight research programmes with missile development. President Nixon's defence secretary, Melvin Laird, has been a strong advocate of letting the Air Force take over the space programme, and during his election campaign Nixon himself hinted that he might share this view. Again, the drastic cuts in the NASA budget by the outgoing Johnson administration could be indications of a general trend in American space intentions: NASA requested $345,000,000 for Apollo and was given $200,000,000. For advanced lunar exploration NASA asked $170,000,000 and received $100,000,000; the orbiting space station allowance was cut from $24,000,000 to $9,000,000. One of the openly admitted functions of NASA was to mount the biggest public relations exercise in history: the United States government may well decide that when it is no longer making prestige headlines with human missions, the time has come to continue space research in the privacy of censorship. Already mild attempts have been made to hold back details of NASA's plans for an orbiting space station.

There are intriguing anomalies in the 1958 National

Aeronautics and Space Act which could provide a loop-hole for drastic changes in American space policy: Section 102A states: 'The Congress hereby declares that it is the policy of the United States that activities in space should be devoted to peaceful purposes for the benefit of all mankind.' However in the same section Clause B states '. . . except that activities peculiar to or primarily associated with the development of weapons systems, military operation, or the defence of the US . . . shall be the responsibility of the Department of Defence.' The outcome will very much depend on the attitude of President Nixon, since the settlement of demarcation disputes between NASA and the Department of Defence is vested in the President.

The prospect of space going military may appal many scientists, but the unpalatable facts are that a large number of advances in space technology were only made to fulfil an immediate military goal. Even Galileo conducted his experiments (demonstrating that the acceleration of gravity was independent of the weight of an object), primarily in order to improve the range tables of a cannon – and the weights he used were large and small cannon balls.

The future of the moon would seem to be as a source of valuable materials, as an observatory, and perhaps as a military base. By the 1970's it may be receiving its first non-astronaut visitors – presumably mainly geologists and astronomers. It seems improbable however, considering the inevitable expense of transport in the foreseeable future, that the science-fiction dream of package holidays to the moon will ever come true. Nevertheless, at a symposium on the commercial uses of space in 1968, a number of tourist proposals were mentioned. Almost predictably a Space Hilton was suggested – a vast orbiting hotel with variable gravity suites. With low gravity the sports potential of lunar games is an amusing field for speculation, but on a more serious level it has already been suggested setting up space hospitals to take advantage of either weightlessness or lunar gravity for helping burns casualties, and remedying other disorders where skin contacts are a problem; however, most experts regard such suggestions as escapist fantasy, for such patients would have to survive the extreme acceleration necessary to rise into space before benefiting from space weightlessness.

The allure of space flight has even sparked Dr von Braun himself to design a passenger-carrying spacecraft which he predicts could be ready in not much more than ten years from now.

Before passengers land on the moon it is almost certain that the focus of public interest will have transferred to the exploration of Mars and Venus. Both the USSR and the United States have already embarked on advanced projects for the unmanned exploration of the planets. Before NASA's budget was cut by the US Congress, it was considering a manned landing on Mars in 1981. Using largely existing Apollo equipment, plus a lifting-body planetary capsule the plan was to put a crew of up to eight men on Mars for ten to forty days.

The Space Science Board of the US National Academy of Sciences is strongly in favour of planetary exploration – but prefers the use of unmanned probes. It suggested that the first American Mars landing be attempted in 1975. Mars gets fairly close to earth once every 780 days, and closest every 16 years. Unmanned probes have already been to Mars, notably Mariner 4, transmitting 22 photographs in July 1965, and Mariner 6 and 7, whose superb pictures of Mars were received in July and August 1969.

In astronomical terms the trip to the moon is just a short hop. An interplanetary journey to Mars or Venus involves much greater feats of three dimensional navigation. Because the planets orbit the sun at different speeds it is not possible to simply aim at a target and fly straight there – by the time any spacecraft reached the point in space where the target was visually sighted, the target would have travelled millions of miles farther along its orbital path. Going to the moon presented similar problems, but on a smaller scale. Earlier this year I asked Dr von Braun in a film interview for the BBC documentary 'Out of this World', if he had ever doubted the future of space travel. He said 'No'. Then he added, 'Let me say this, in retrospect, with all the advantage of twenty twenty hindsight, I sometimes wonder at the naïveté that I myself and many of my associates had in the earlier days. For example the problem of navigating to the moon and making a pinpoint landing where you desire looks awfully simple in a motion picture. But when you have to do it, and you consider that the moon is moving, and the earth is spinning, there are very difficult celestial mechanics involved – it's a very, very difficult problem. I ask myself sometimes how we ever hoped to solve these problems without the help of the fantastic computation machines that we have today.' On a moon voyage the gravitational fields of the earth and the moon are a considerable navigational problem – but on an interplanetary voyage the pull of the sun and other planets has to be taken into account. Because of this, curious anomalies crop up when man considers going to Mars: with a velocity of seven miles a second it would take over 250 days to reach the Red Planet – but speed the trip up to ten miles a second and it only takes 70 days. With an initial velocity of seven miles a second it takes 146 days to reach Venus. The Russians hit it with Venus 3, and Venus 4, launched in June 1967, succeeded in achieving a soft landing. If the Venus 4 robot had been able to scan the Venusian sky two days after its touch down, it would probably have seen America's Mariner 5 flying past. Expeditions to Mars are already viewed as being of much greater scientific importance than the conquest of the moon, and in not very many years it may be difficult to understand how many people alive today were so excited by the thought of lunar exploration.

Until forms of propulsion superior to rocket power have been perfected it is unlikely that man will be able to reach farther than the closest planets: the snag lies in the nature of chemical rockets – there are absolute limits to their exhaust velocity, and therefore to their maximum speed. The Saturn V moon rocket has a combustion chamber temperature in excess of 5,400 deg. F., and it is highly improbable that any future rocket design will make a significant improvement on this key figure: this combustion temperature directly relates to the rocket's exhaust velocity, and if it cannot be improved, chemical rockets will never travel faster.

To overcome this difficulty rocket engineers have been searching for many years for alternatives to chemical combustion as a propellant. As Professor Mueller noted, many physicists have hoped that nuclear energy will provide the answer. One nuclear rocket already being studied uses a small uranium reactor to heat the combustion chamber; instead of burning the hydrogen in oxygen, the reactor heats it to even higher temperatures, giving an improved exhaust velocity. Work on this rocket began in 1955 as a USAF project, and it has now become the joint responsibility of NASA and the Atomic Energy Commission. It produces a 50,000 lb. thrust, with twice the fuel economy of the existing upper stages of the Saturn V. Dr von Braun, the father of the chemical moon rocket, told me that he considered it a highly promising engine. He explained, 'No one envisages flying a rocket, with this nuclear engine actively operating, out of Cape Kennedy. We would put a stage powered by that engine on the top of the Saturn V and fly it with chemical power first into earth orbit. The reason is simply this: a nuclear rocket engine is very harmless, even if you smash it up due to a mishap on the pad, as long as it has never been operated. But once you fire up a nuclear rocket engine and then smash it you have a very major contamination problem on your hands. So we feel far more comfortable firing these things only out of Earth orbit. We believe nuclear rockets will have a great future when man wants to fly to Mars, or when we would like to have an economical supply shuttle-operation to the moon. But nuclear power will always be used from one orbit to another, rather than for going down or taking off from Earth or even landing on the Moon.'

Six weeks before the Apollo moon landing the Americans passed another milestone in the conquest of space. Like Goddard's first flight of a liquid-fuel rocket, it received scanty attention from the Press – the forthcoming moon flight was already filling every space headline. On 11 June 1969 at the Nuclear Rocket Development Station on Jackass Flats in Nevada, the experimental nuclear rocket engine called Nerva was brought up to full power for the first time. In a simulated flight configuration it was fired for thirteen minutes; nuclear scientists watched anxiously as it built up its thrust, producing the desired 50,000 lb. for three and a half minutes. On a previous test a nuclear engine generated so much heat that it completely melted, but now scientists seem confident that they have developed materials capable of withstanding the colossal temperatures created by nuclear reactions. The test on 11 June seemed to prove they have succeeded. Such nuclear testing tends to be a closely guarded secret, and it may be some time before we hear positive proposals for an American nuclear powered spaceflight. The Russians are also known to be pioneering nuclear propulsion, but no details have so far been released. The problem with nuclear-heated rockets is, as with chemical rockets, temperature in the expansion chamber: temperatures of up to 6,000 deg. F. can be withstood by the highest melting-point materials at present known (hafnium carbide and tantalum carbide): such exotic materials are very expensive and no great improvement can be predicted. In theory gaseous core reactors, where the fuel is directly exposed to the nuclear reaction, could achieve temperatures of 18,000 deg. F. – but similar heat problems would be encountered in designing exhaust nozzles; also it is still debatable whether a radioactive exhaust will ever actually be permitted, even in the depths of outer space. As Dr von Braun hinted, atomic rockets could be unpopular because of the danger, however remote, of their nuclear reactors returning from space to crash in populated areas at colossal speed. Moreover, for manned space flight their weight advantages (they do not carry liquid oxygen) are largely lost because of the heavy shielding needed to protect the astronauts from radiation.

One of the more extreme suggestions for using nuclear power came from a Princeton Professor: he recently suggested a 40,000,000-ton rocket, propelled by 30,000,000 hydrogen bombs; a bomb would be dropped and exploded at the tail end of the vehicle every 100 seconds, accelerating a capsule containing all the inhabitants of Princeton to a speed of 600 miles a second, and committing them to an intergalactic tour lasting several millennia. Such a proposal sheds more light on the Professor's hopes for the inhabitants of Princeton than on intergalactic research, but other experts have been known to discuss such methods of propulsion quite seriously.

Meanwhile, physicists are developing propellants that will not be restricted by the 'temperature barrier'. Exhaust velocity is of primary importance, and many believe that the rocket of the future will be driven by the ejection of a stream of electrically charged particles of sub-microscopic size, but extremely high speed. Acceleration would be slow, but it could be sustained for a very long time. In space such vehicles could eventually build up colossal velocities.

One feasible type of 'electric rocket' is the ion rocket. An ion is what is left of an atom after an electron has been removed. The electron takes its negative electric charge with it, leaving the ion positively charged. If an atom is 'ionised' between two plates with a voltage difference, the electron will rush towards the positive plate, and the ion towards the negative plate. This, very roughly, is the principle of the ion rocket. Caesium vapour is ionised at a temperature of 1,800 deg. F. in a chamber with a very high voltage across it: the 'negative plate' becomes the back of the rocket and, as the ions rush towards it they pass through a hole in it, and emerge to produce the required propulsive effect. A little caesium 'fuel' goes a long way, as the mass loss is very small. However a considerable amount of electrical energy is required to accelerate the ions and provide the ionising temperature. Some experts suggest overcoming this problem by using a small nuclear reactor.

The plasma rocket is a more sophisticated form of propulsion of the same basic type. Plasma is a mixture of electrons and dissociated ions at very high temperature. A combination of electric and magnetic fields causes plasma to accelerate to velocities approaching the speed of light, thus making it a very promising rocket 'fuel'.

Ion and plasma motors have already been designed, and both Goddard (in 1906) and Oberth (1929) considered the possibility of ion propulsion. Tests have been promising, but no one has so far succeeded in producing a thrust of more than a fraction of an ounce. An experimental ion engine was used for attitude control on Voskhod 1, and a plasma jet on the Zond 2 Mars probe.

For even greater speeds some researchers have even suggested using light itself as a power source: it travels at 186,282 miles a second, and theoretically the ultimate in rockets can only be powered by an actual beam of light. Although light has no mass and therefore can generate no thrust, it can be considered as a stream of energy pulses, called photons. Photons, although possessing no measurable mass, exert a measurable pressure. The pressure of photons in the sunlight falling on earth is about 2 millionths of a pound per square yard. After launching a spaceship in a conventional manner, one idea for photon propulsion is to erect vast sails several square miles in area, facing the sun: in theory photon pressure would slowly accelerate the ship to the speed of light. The fault in this idea is, however, that at mid-point between our sun and another star the pressures from both sources would be equal, and the spacecraft becalmed. Other theories of photon propulsion still depend on the production of a substance which, by definition, does not exist: it is called anti-matter, and it annihilates matter on contact, releasing radiant energy which could propel the photon rocket. Such ideas are as obscure as they sound, but it is only by using just such an improbable source of power which today exists purely in theory, that man could reach even the nearest star, Alpha Centauri, 4·3 light years away.

In theory a photon rocket could reach the speed of light after a year's acceleration, and visit stars within 25 light years' range to return the rocket's occupants at the end of their lives. However as the edge of our galaxy is 20,000 light years away, and the first neighbouring galaxy an additional 140,000 light years distant, the Princeton Professor's dreams of intergalactic travel must remain in the realms of fantasy. Also, according to Einstein's theory of relativity, as any rocket approached the speed of light it would grow bigger and bigger until it filled the universe! Such concepts sound like overgrown fairy-tales, but one may recollect that until recently no one knew of the existence of quasars, sources of radio waves already thousands of millions of light years away, and apparently continuing their relativistic velocities of recession at more than 80% of the speed of light.

In comparison to such speeds and distances, with the nearest star in our galaxy being 4·3 light years away, the moon suddenly seems a very close neighbour. By the end of this year, 1969, it will reveal many more of its secrets, and inquisitive scientists may well begin to look elsewhere for new challenges as lunar travel becomes routine. Perhaps a sense of international anti-climax will ensue, for the challenge of Mars is more obscure: everyone has seen the moon, but not so many can recognise Mars in the night sky.

As the Saturn V, carrying the Apollo 11 team to walk on the moon, soared into the sky above Cape Kennedy on 16 July 1969 it was watched by two of the original pioneers. One was 76-year-old Hermann Oberth – he had flown from his home in Europe specially to see the launch of the mission which converted his boyhood fantasies into a historical fact. The other was Wernher von Braun – today a sprightly, dynamic 57-year-old American citizen, and Director of the Marshall Space Flight Centre. Few men can ever have lived with such an ambitious dream and seen it become reality. And for Dr von Braun, it has been an even more incredible story – he is not only honoured in a former enemy country, but he has succeeded in persuading that country to pay for the realisation of his dream. I asked him if he remembered how his interest in rocketry began. He replied, 'With me it started with the moon, believe it or not. I was thirteen then, a high school student. I was very interested in astronomy in the amateur sense. I liked to look at the moon and the stars, and when I had my Lutheran confirmation, my parents gave me a telescope, and that – that was like giving a boy a motorbicycle, I guess, and my interest in astronomy has never faded away. It was actually in one of those amateur astronomer magazines I had at that time, 1925, that I stumbled across an advertisement of a book written by Professor Hermann Oberth. It was entitled *The Rocket to Planetary Spaces*, and I sent for that book: of course I couldn't understand a word, because it was full of mathematical formulae and I was just a schoolboy. But that – that really kicked me off! You know, when you are thirteen years old the world looks awfully simple and beautiful, and when I read this book – although I didn't understand all of it – I got the message loud and clear that people can fly one day with rockets. Not only to the moon, but also to the planets. I decided right then and there, that it was for me – I just fell for the cause!'

The main history of that 'cause' spans less than fifty years – yet, as we have seen, it has been an extraordinary story of triumph and disaster, of technological miracles, death, and the fulfilment of an impossible dream. It has been an international story involving many nations and thousands of people. It is therefore perhaps unfair, if not impossible, to single out one individual as being the inspiration of the conquest of the moon. Certainly military and political considerations have played a major role in pushing both the United States and Russia into the space race. Ex-President Lyndon Johnson deserves more credit than generally acknowledged, for, as Vice-President, he was the man who probably did most to persuade President Kennedy to commit the United States to a moon landing deadline.

Ironically, however, the man perhaps more responsible than any other in recent years for the American achievement is Wernher von Braun – the German who might have been lynched if he had faced a crowd of Londoners at the end of the last war. He himself insists that the success of Apollo was the result of good teamwork. I asked Dr Ernst Stuhlinger, a member of von Braun's team from the early days, and today Associate Director for Science at the Marshall Space Flight Centre, what part he felt Dr von Braun had played in the conquest of the moon. He replied, 'He has been the great moving dynamo and spirit behind it all the time. There's no doubt about that. He has always filled his team, not only with ideas and with incentive, but also with hope and faith and with the real strong belief in the sense of what we were doing. He certainly deserves all the credit for this development.' Many, including Dr von Braun, might disagree. Nevertheless he is in a unique position today, having devoted a lifetime to rockets and developed them from little more than sophisticated fireworks to his notorious V-2 and finally to the Saturn V. In this Chapter some of the more advanced projects for the exploration of Space have been reviewed. I asked Dr von Braun to prophesy the more immediate prospects for space travel beyond the moon in the next few years. He replied with the confidence of a man whose prophecies have proved right before: 'My personal guess will be that for a number of years we'll continue to explore the planets with unmanned equipment. We did the same thing with the moon – fly-bys, orbiters, unmanned landers – to get as much information on these planets as possible. We will not limit this solely to the nearer planets like Venus and Mars, but also we'll take an interest in the outer planets, Jupiter and Saturn – we know very little about them. Man will then probably follow to a few of these planets wherever he finds the environment interesting enough, and he may use nuclear rockets to get there: that, I believe, is anywhere between one and two decades away – a manned flight to Mars.'

The phrase 'Crying for the moon' as a symbol of reaching out for the unattainable has become outdated, and the future of manned space flight towards more distant goals remains uncertain. As Shakespeare said:

> The fortune of us that are but Moon's men
> doth ebb and flow like the sea.
> (Henry IV, Part I.)

Q

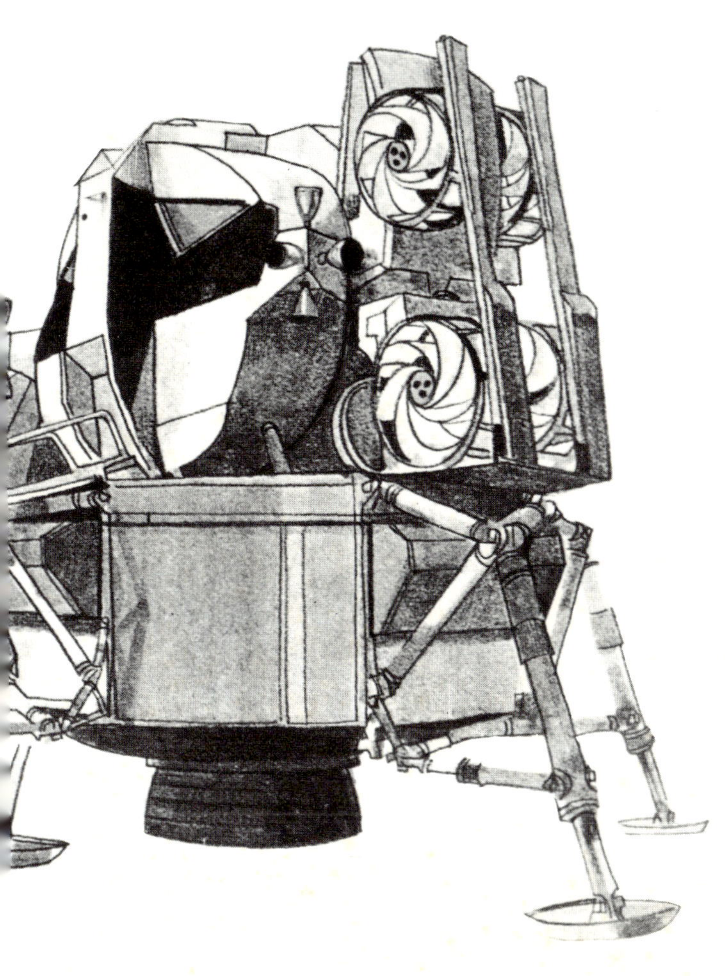

Above *An artist's impression of Grumman's Lunar Payload Module, carrying a lunar vehicle on its side.*

Above *An artist's impression of some lunar mobility aids designed by Grumman for NASA.*
Right *The flexible wheels are designed to 'flow' over obstacles.*

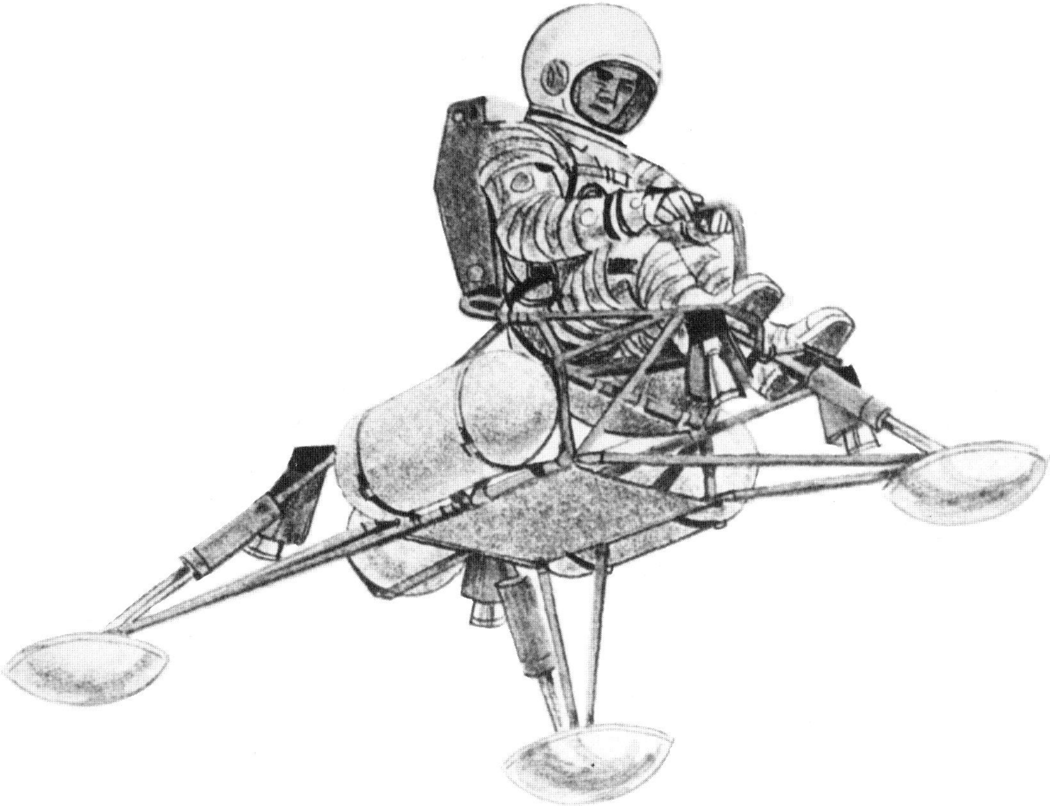

Above *Grumman's Lunar Roving Vehicle, designed to be stowed in the descent stage of the lunar module. An alternative six-wheeler would be unmanned, and remotely controlled from earth.*

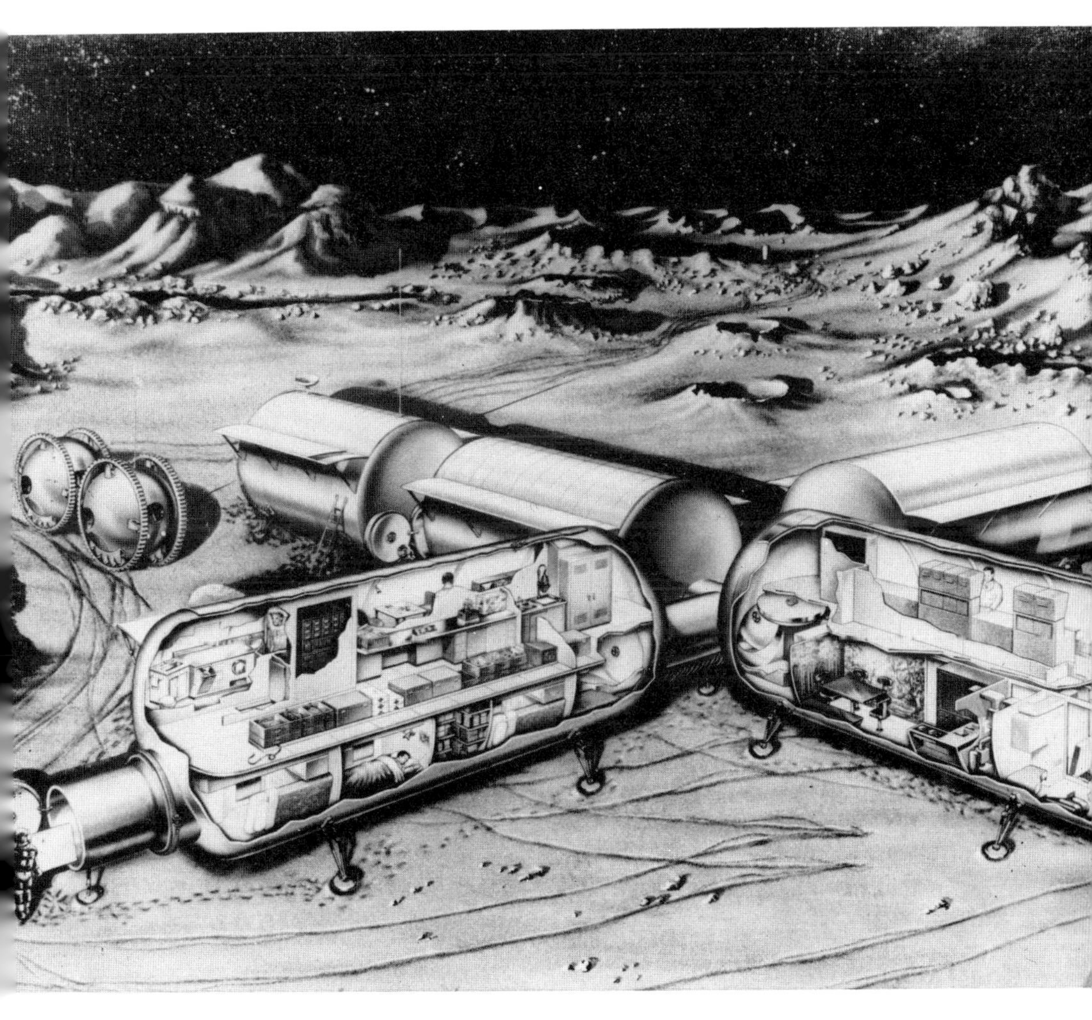

Above *One concept of a moon base prepared by the Lockheed Company for NASA: Electrical power is supplied by a remote nuclear power source.*

Milestones of the Race to the Moon

		USSR	USA
4 Nov	1957	**Sputnik 1** First artificial earth satellite	
3 Dec	1957	**Sputnik 2** First biological study satellite	
1 Feb	1958		**Explorer 1** First detection of Van Allen belts
2 Jan	1959	**Lunik 1B** First 'freedom' from earth – missed moon by 4,600 miles; objective unknown	
3 Mar	1959		**Pioneer 4** Missed moon by 37,000 miles instead of intended 15,000 miles
7 Aug	1959		**Explorer 6** First TV pictures of the earth
12 Sept	1959	**Lunik 2C** Impacted on moon	
4 Oct	1959	**Lunik 3D** Circumlunar flight; first pictures of hidden face of the moon	
1 April	1960		**Tiros 1** First meteorological satellite
13 April	1960		**Transit 1B** First navigational satellite
10 Aug	1960		**Discoverer 13** First recovery of a space capsule
12 Aug	1960		**Echo 1** First permanent telecommunications satellite
19 Aug	1960	**Sputnik 5** First recovery of a capsule containing live animals	
12 April	1961	**Vostok 1** First human orbital flight: Gagarin	
7 Mar	1962		**OSO 1** First satellite observatory
23 April	1962		**Ranger 4** Impacted on rear of moon. Failed to take photographs
12 Aug	1962	**Vostok 3** and **4** First meeting of manned spacecraft in space	
18 Oct	1962		**Ranger 5** Missed moon by 450 miles; identical mission to Ranger 4
16 June	1963	**Vostok 6** First woman in space (Valentina Tereshkova)	

		USSR	USA
26 July	1963		**Syncom 2** First satellite kept on a synchronised orbit
28 Sept	1963		**Transit 5B** First satellite entirely fuelled by generated nuclear isotope
17 Oct	1963		**Vela 1** and **2** First satellites to detect nuclear explosions
30 Jan	1964		**Ranger 6** Impacted on moon – failed in photo mission
28 July	1964		**Ranger 7** First pictures close to surface of moon – returning 4,300 photos
12 Oct	1964	**Voskhod 1** First multiple space ship	
30 Nov	1964	**Zond 2** First use of a plasma rocket	
17 Feb	1965		**Ranger 8** Returned 7,100 pictures of lunar surface
18 Mar	1965	**Voskhod 2** First walk in space: Leonov	
21 Mar	1965		**Ranger 9** Returned 5,300 pictures of lunar surface
23 Mar	1965		**Gemini 3** First changing of the orbit of a manned spaceship
6 April	1965		**Early Bird 1** First commercial satellite for telecommunications
9 May	1965	**Lunik 5** Impacted on moon – failed to soft-land	
3 June	1965		**Gemini 4** First manned propulsion outside craft
8 June	1965	**Lunik 6** Bypassed moon	
4 Oct	1965	**Lunik 7** Impacted on moon; failed to soft-land	
3 Dec	1965	**Lunik 8** Impacted on moon; failed to soft-land	
4 Dec 15 Dec	1965 1965		**Gemini 7** and **6** First rendezvous in space
3 Feb	1966	**Lunik 9** First soft-landing on moon and first pictures taken on moon	
16 Mar	1966		**Agena** and **Gemini 7** First meeting of space vehicles in orbit

		USSR	USA
31 Mar	1966	**Lunik 10** First artificial moon satellite—achieved lunar orbit 3 April; measured magnetic field, gamma rays	
30 May	1966		**Surveyor 1** Successful soft-landing on lunar surfac televised photographs to earth
10 Aug	1966		**Lunar Orbiter 1** First pictures taken by a satellite of the moon
19 Nov April	1967 1968	**Cosmos** Automatic rendezvous and docking of two unmanned satellites	
Jan	1968		**Surveyor 7** First sampling of lunar surface
Sept	1968	**Zond 5** First flight round moon, back to earth and recovery, unmanned	
Dec	1968		**Apollo 8** First manned mission to leave earth's influence and orbit the moon
16 Jan	1969	**Soyuz** First docking by two manned spacecraft	
May	1969		**Apollo 10** First manned trip to within 10 miles of lunar surface. First manned link-up and transfer of crews in lunar orbit
July	1969		**Apollo 11** Man on the Moor

Recommended Additional Reading

1 **'A Handbook of Astronautics'**
Edited by S. W. Smith
University of London Press 1966; Dufour Editions 1969

2 **'Advances in Space Science and Technology'**
Edited by F. Ordway
Academic Press 1967

3 **'Apollo and the Universe'**
Edited by S. T. Butler and H. Messel
Pergamon 1968

4 **'Assault on the Moon'**
Eric Burgess
Hodder & Stoughton 1966

5 **'Astronautics in the Sixties'**
K. W. Gatland
Iliffe 1963

6 **'Exploration of the Moon'**
F. Branley
Nelson 1965; Doubleday 1966

7 **Handbook of Soviet Space-Science Research**
G. E. Wukelic
Gordon & Breach 1968

8 **'History of Rocketry and Space Travel'**
W. von Braun and F. Ordway
Nelson 1967; Thomas Y. Crowell 1966

9 **'Inventing the Future'**
D. Gabor
Secker & Warburg 1963; Knopf 1964

10 **'Manned Satellites'**
W. F. Hilton
Hutchinson 1965; Harper & Row 1966

11 **'Space Science and Engineering'**
Edited by E. Stuhlinger and G. Mesmer
McGraw-Hill 1965

12 **'Spaceflight Today'**
K. W. Gatland
Iliffe 1964; Aero 1963

13 **'Survey of the Moon'**
P. Moore
Eyre & Spottiswoode 1963; W. W. Norton 1963

14 **'The Encyclopaedia of Space'**
Paul Hamlyn 1968; McGraw Hill 1968

15 **'Venture into Space'**
A. Rosenthal
NASA 1968

16 **'Virgil "Gus" Grissom'**
The Macmillan Company 1968

Index